T0259720

SpringerBriefs in Applied Sciences and Technology

SpringerBriefs present concise summaries of cutting-edge research and practical applications across a wide spectrum of fields. Featuring compact volumes of 50–125 pages, the series covers a range of content from professional to academic.

Typical publications can be:

- A timely report of state-of-the art methods
- An introduction to or a manual for the application of mathematical or computer techniques
- A bridge between new research results, as published in journal articles
- A snapshot of a hot or emerging topic
- An in-depth case study
- A presentation of core concepts that students must understand in order to make independent contributions

SpringerBriefs are characterized by fast, global electronic dissemination, standard publishing contracts, standardized manuscript preparation and formatting guidelines, and expedited production schedules.

On the one hand, **SpringerBriefs in Applied Sciences and Technology** are devoted to the publication of fundamentals and applications within the different classical engineering disciplines as well as in interdisciplinary fields that recently emerged between these areas. On the other hand, as the boundary separating fundamental research and applied technology is more and more dissolving, this series is particularly open to trans-disciplinary topics between fundamental science and engineering.

Indexed by EI-Compendex, SCOPUS and Springerlink.

More information about this series at http://www.springer.com/series/8884

Shaharin Anwar Sulaiman
Editor

Sustainable Thermal Power Resources Through Future Engineering

Springer

Editor
Shaharin Anwar Sulaiman
Department of Mechanical Engineering
Universiti Teknologi Petronas
Seri Iskandar, Perak, Malaysia

ISSN 2191-530X   ISSN 2191-5318   (electronic)
SpringerBriefs in Applied Sciences and Technology
ISBN 978-981-13-2967-8   ISBN 978-981-13-2968-5   (eBook)
https://doi.org/10.1007/978-981-13-2968-5

Library of Congress Control Number: 2018958493

This Springer imprint is published by the registered company Springer Nature Singapore Pte Ltd.
The registered company address is: 152 Beach Road, #21-01/04 Gateway East, Singapore 189721, Singapore

Preface

Thermal power is a product involving energy conversion process with practical coverage of thermal and fluid components and systems. It involves complex knowledge in thermodynamics, heat transfer and fluid mechanics, which is usually mastered by graduates in mechanical engineering. In electric power plants, thermal power serves as the primary product before electricity can be generated. Although widely related to electric power plants, the knowledge in thermal power is also widely applied in areas such as air conditioning, internal combustion engines, and various industrial applications. Hence, the issues of energy efficiency and environment are critical in the production of thermal power. Various and diverse research works have been carried out worldwide in order to improve aspects in usage and production of thermal power. The biggest leap in improvement of thermal power probably occurred after the 1973 oil crisis, which saw a limited production of oil in the global market, resulting in sudden increase in energy prices. On another aspect, the use of air conditioners in homes and offices around the world has been the main driver of global energy demand over the next three decades; this leads to urgent need to improve cooling efficiency. The emergence of Fourth Industrial Revolution would, in addition, result in increased demand in the capacity of data centers. This would not happen without accompanying increase in demand of cooling energy for data centers. Motivated by these issues, this book shares the efforts by researchers in science and engineering on conventional and renewable energy, and energy efficiency vis-à-vis thermal power. The editor would like to express his gratitude to all the contributing authors for their effort in preparing the manuscripts for the book. May it serve as a useful reference to readers.

Seri Iskandar, Malaysia   Shaharin Anwar Sulaiman

Contents

Experimental Investigation on Performance of Solar-Powered Attic Ventilation

Firdaus Basrawi, Thamir K. Ibrahim, Surendran S. Sathiyaseelan and A. A. Razak

An attic ventilation system has the potential to reduce cooling load and power consumption of an air-conditioning system in buildings that experience elevated temperature in the attic space during the day. For this, a solar-powered attic ventilation system, which harnesses the abundant source of solar energy from the sun, is desired to ventilate the hot attic space in an economical way. On the other hand, it is known that the efficiency of photovoltaic (PV) cells decreases when temperature of the solar panel increases. This chapter discusses the design and development of an efficient and low-cost solar-powered attic ventilation system. The novelty approach is intended to improve the PV efficiency and the overall performance of the system by providing airflow for the solar panel from outflow of the ventilation system. The designed and developed solar-powered attic ventilation system consists of an exhaust fan powered by a polycrystalline solar panel, a direct current to direct current (DC-DC) step-down converter power module, ducting system, and ventilation casing assembled together. Components like ventilation casing and angle adjustable mounting for solar panel are prepared through fabrication work. A few experiments and testing on ventilation process and PV efficiency are conducted to determine the performance, working condition, and functionality of the developed system. Parameters like ambient temperature and indoor temperature of the research location are studied in the experiment of ventilation process. Solar irradiance, the power produced from the PV, and the power delivered to the exhaust fan are studied through the experiment on PV efficiency. The developed solar-powered attic ventilation system reduces the attic temperature by 2.9 °C and keeps the temperature difference between the ambient and indoor in the range of 0.1–0.4 °C. An increment of about 17% was observed for the PV efficiency when there is airflow for the PV module from outflow of the ventilation system.

F. Basrawi (✉) · T. K. Ibrahim · S. S. Sathiyaseelan · A. A. Razak
Universiti Malaysia Pahang, Pahang, Malaysia
e-mail: mfirdausb@ump.edu.my

T. K. Ibrahim
Tikrit University, Tikrit, Iraq

© The Author(s), under exclusive license to Springer Nature Singapore Pte Ltd. 2019
S. A. Sulaiman (ed.), *Sustainable Thermal Power Resources Through Future Engineering*, SpringerBriefs in Applied Sciences and Technology,
https://doi.org/10.1007/978-981-13-2968-5_1

Introduction

The Energy Commission of Malaysia shows that the electricity demand in Malaysia increased from 18,882 MW in 2013 to 19,845 MW in 2014. The residential sector consumes about 21% of the energy supply International Energy Agency. Approximately 20% of this portion is used to power air conditioning [1, 2]. The total number of air-conditioning units in residential buildings owned by Malaysians was 582,792 in 2000 [3]. It was also estimated that this figure would increase to 726,504 units in 2005 and by roughly 1,217,746 units in 2015 [3]. In another report [4], it was estimated that about 75% of Malaysians were depending on air conditioning for a better thermal comfort.

However, the widespread use of air-conditioning system in homes and buildings is quite unsatisfying [1]. This is because an air conditioner consumes a large amount of energy and incurs high electricity bill to consumers [5]. Moreover, the problem of higher energy consumptions will arise if the air conditioner is run throughout a hot day. During a hot day, heat builds up in the attic space because of incident solar energy on the roof. The heat gets trapped within the attic space and is transferred from the attic floor to the space below it, thus increasing the indoor temperature. Consequently, the cooling load of the air-conditioning system will increase. The air-conditioning system must then operate longer to reduce the temperature and consume more power to overcome the extra thermal load in the building.

Therefore, there is a need of ventilation system to extract out hot air from the attic space and to reduce the power consumption of air conditioners. There are few ventilation methods like natural ventilation and mechanical ventilation which have been early efforts in minimizing the effects of heat buildup in the attic space. Currently, the most commercially available ventilation system is the turbine ventilator system. The American Society of Heating, Refrigerating and Air-Conditioning Engineers (ASHRAE) [6] defines the turbine ventilator as a heat escape gate located on the roof and is suitably enclosed for different weather conditions, with the major motive forces being wind induction and the stack effect.

The turbine ventilator system, however, is dependent on the wind direction, wind speed, and the stack effect. If the speed of the wind is low, the turbine ventilator could not work effectively to ventilate the attic space efficiently in order to displace the hot air. Additionally, the turbine ventilator system is quite costly for a low airflow rate system. Thus, these systems have proven to be less efficient and economical. Therefore, what is needed is an improved ventilation system that will properly ventilate the attic space and reduces the cooling load of air conditioner. Together, it would be applicable to many roof configurations with low operation and installation costs relative to other ventilation systems.

Thereby, solar-powered attic ventilation system is discovered as a new alternative to enhance ventilation process in the attic space. Solar-powered attic ventilation system harnesses the abundant source of solar energy from the sun to ventilate out the hot air in the attic space in a more economical way [7]. It consists of solar panel that absorbs the sunlight and converts it into electrical energy which powers the exhaust fan [8]. The proposed solar-powered ventilation system is economically low-cost, significantly reduces the cooling load and energy consumption, requires minimum maintenance, and is environment friendly.

Meanwhile, the use of solar panel in the proposed ventilation system leads to the interest in the efficiency of solar panel as well. It is known from the studies that the efficiency of the solar panel will decrease if the temperature of the photovoltaic (PV) panel increases [9]. There are some passive and active cooling systems that are readily available to maintain the efficiency of the PV panels [10]. Despite that, this research is also intended to explore a novel cooling system for PV panels. A passive cooling system in which requires no input energy system is proposed, in which the air ventilated from the attic space is utilized for cooling of the PV panel. The ventilated air is directly channeled to the back surface of the PV panel via a ventilation and ducting system.

The work presented in this chapter aims to design and develop a solar-powered attic ventilation system that will facilitate the ventilation process of the attic space efficiently and effectively. At the same time, the solar-powered ventilation system is integrated with PV panel cooling system that will benefit each system concurrently under one consolidated system.

Methodology

Design and Development Process

Figure 1 shows an overall view of the solar-powered ventilation system and how each component is related to each other in working as system. It consists of an exhaust fan that is powered by a solar panel, ventilation system, and ducting system. The exhaust fan is assembled inside the research location while the solar panel is mounted on an angle adjustable mounting structure which is placed on the rooftop of the research location. The transmission of power from the solar panel to the exhaust fan is via a simple electrical circuit that contains wiring system, switch, and direct current to direct current (DC-DC) step down converter power module. The ventilation system includes of a box casing with shutter sealing, whereas the ducting system is made of a flexible ducting pipe that connects between the exhaust end of the exhaust fan and the back surface of the ventilation casing.

The design and development process of the solar-powered attic ventilation system begins with the formulation and calculation for the problem of the project. Firstly, the size or the space volume of the research location is determined. The research location is a cabin container in the research site of the Energy Sustainability Focus Group of Faculty of Mechanical Engineering of Universiti Malaysia Pahang, Pekan Campus. The dimension of the cabin container is 20 ft × 8 ft × 8.6 ft with a volume of 1376 ft^3.

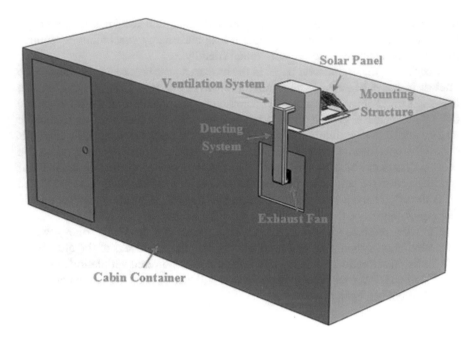

Fig. 1 Helicopter view of the solar-powered ventilation system and its component

Table 1 Specification of exhaust fan	Power	15 W
	Voltage	12 V
	Current type	Direct current (DC)
	CFM	300 CFM
	RPM	1450 RPM

Next, the airflow rate or CFM—cubic feet per minute—of the exhaust fan to be used to ventilate the hot air out from the cabin container is calculated based on the indoor volume space of the cabin container. The air changes per hour (ACH) for the cabin container is assumed to be 10 (ASHRAE Standards for Ventilation is referred). The CFM calculation for the cabin and the exhaust fan is:

$$CFM = \frac{Volume\ of\ room \times ACH}{60} \tag{1}$$

Thus, the calculated cabin container CFM is about 229.33 CFM. Hence, a suitable exhaust fan with a CFM range of 200–300 CFM or higher is considered for this research. Subsequently, the voltage and power rating of the exhaust fan are taken into account for determining the suitable power rating for the solar panel. Table 1 shows the specification of the exhaust fan, and Table 2 shows the specification of the solar panel used in the research.

Table 2 Specification of solar panel

Maximum power	10 W $\pm$ 3%
Voltage at maximum power	17.28 V
Current at maximum power	0.58 A
Open circuit voltage	21.42 V
Short circuit current	0.66 A
Maximum system voltage	1000 V
PV type	Polycrystalline
PV surface area	0.098 m^2

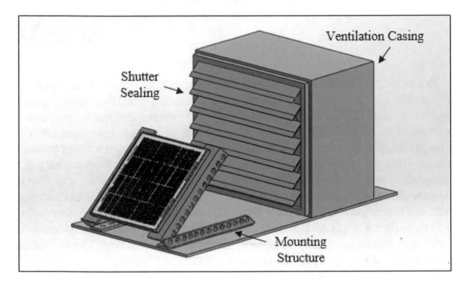

Fig. 2 CAD of solar-powered attic ventilation system

After determining the sizing and dimension of the exhaust fan and solar panel, the designing process of the solar-powered ventilation system is carried out. It is also proceeded after some detailed reviews of previous art and patents. This is to ensure the newly modeled design is different from the existing design.

Figure 2 shows the CAD of the solar-powered attic ventilation system. The desired model of the solar-powered ventilation system is designed using computer-aided drawing (CAD) software, solid works. It consists of a ventilation casing fitted with shutter system that faces the back surface of the solar panel. The louver for the shutter system is designed in such a way for the proper sealing of the ventilation system to prevent the entry of water and small insects. The mounting structure is to place the solar panel on top it and is designed on the concept that its angle can be varied between 0 and 90°. The initial CAD design, however, might look different from the fabricated model due to the availability of material and limitation of the research location, where there is no attic space in the cabin container.

Fig. 3 Ventilation casing

The fabrication work in this project is mainly to fabricate the ventilation casing with shutter and the angle adjustable mounting structure for the solar panel. Figure 3 shows the fabricated ventilation casing. The raw material used to fabricate both the ventilation casing and the mounting structure is galvanized iron sheet metal of 1 mm thickness. The fabrication process includes the shearing (cutting) process, bending process, filling, grinding, and drilling process. The ventilation casing is a cube-shaped box that fits a shutter with openings at the front surface for airflow to the surrounding environment and has a hole of 11 cm diameter at the back surface to fit and locate the flexible ducting pipe. The dimension of the ventilation casing is 31 cm × 31 cm × 31 cm. The inner part of the ventilation casing is insulated with rubber-type insulation material to absorb heat from the surrounding. The ventilation casing is also sealed tightly to prevent entry of water and insects.

Figure 4 shows the angle adjustable mounting structure for three different angles: 5°, 45°, and 90°. The angle adjustable mounting structure for the solar panel is stand structure that places the solar panel and can be tilted to various angles in the range 0°–90°. The panel mounting is designed in such way that it is angle adjustable to make it useful for various shapes of attic or roof with various angles of inclination. It consists of two L-shaped metal bar on top and bottom parts, being supported by a metal stand in between them on each side of the structure. Each of it has a dimension of 20 cm × 3 cm × 0.1 cm.

Once the fabrication process complete, all the components can be assembled together to complete the process of designing and developing of the solar-powered attic ventilation system. Figure 5 shows the assembly of the ventilation casing and the mounting structure with solar panel, and Fig. 6 shows the assembly of the ducting pipe to the exhaust fan. The solar panel is first placed on the angle adjustable mounting structure. Next, the solar panel with the mounting structure and the ventilation casing is placed on the rooftop of the research location. The flexible aluminum ducting is then inserted into the hole made on the back surface of the ventilation casing and is sealed nicely. The other end of the ducting is to be connected to the exhaust side of the exhaust fan, which is assembled at the mirror side inside the cabin container.

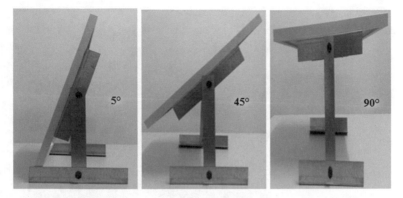

Fig. 4 Angle adjustable mounting structure

Fig. 5 Assembly on the top of roof of cabin container

Figure 7 shows the assembly of the exhaust fan and the wiring system inside the cabin container. The exhaust fan is assembled inside the cabin container and the wiring system is done from the solar panel through the DC-DC adjustable step-down converter power module, switch, and to the exhaust fan. The assembly process is therefore complete. Figure 8 shows the final product of the designed and developed solar-powered ventilation system.

Experimental Setup

The designed and developed solar-powered attic ventilation system will be used to conduct several experiments to test the overall performance and the efficiency of the developed system. There are altogether four different testing methods to be carried out under three different experimental setups. Table 3 shows the testing methods.

Fig. 6 Assembly of the
ducting pipe to the exhaust
fan

Fig. 7 Assembly of the
exhaust fan and wiring
system

Figure 9 shows the experimental setup for natural ventilation (Test 1) while Fig. 10 shows the experimental setup for solar-powered ventilation (Test 2). It can be seen from Fig. 9 that the sliding window is half opened and an approximate area of $0.40\,\text{m}^2$ is allowed for the natural ventilation to take place.

Figure 11 shows the experimental setup for Test 3, where PV module is placed close to the ventilation system. Figure 12 shows the experimental setup for Test 4, where PV module is placed far from the ventilation system.

Fig. 8 Final product of the solar-powered ventilation system

Table 3 Testing methods

Test	Parameter observed
1. Natural ventilation	Ambient temperature, indoor temperature of the cabin container, and solar irradiance
2. Solar-powered ventilation	Ambient temperature, indoor temperature of the cabin container, and solar irradiance
3. PV module efficiency with cooling effect	Ambient temperature, input and output voltage, input and output current, and solar irradiance
4. PV module efficiency without cooling effect	Ambient temperature, input and output voltage, input and output current, and solar irradiance

Fig. 9 Experimental setup for natural ventilation

Fig. 10 Experimental setup for solar-powered ventilation

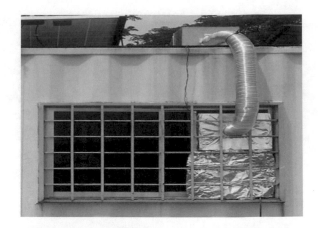

Fig. 11 Experimental setup for module close to the ventilation system

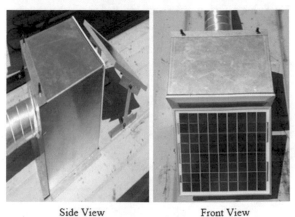

Side View   Front View

Fig. 12 Experimental setup for module far from the ventilation system

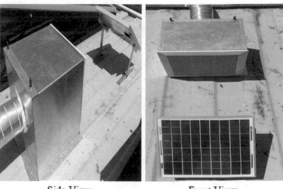

Side View   Front View

Data Collection and Analysis

The data collection process is carried out for all the testing after the experimental setup for each test is complete. For Test 1, the data for the ambient temperature, indoor temperature of the cabin container, and irradiance are collected. These data are collected for every one hour from 8.00 am to 6.00 pm for three consecutive days (April 28–30, 2017). The data collection process for Test 2 and Test 3 are to be carried out simultaneously under one experimental setup. Test 2 involves the data collection of the ambient temperature, indoor temperature of the cabin container, and irradiance for every one hour from 8.00 am to 6.00 pm for three consecutive days, while Test 3 involves the data collection for voltage and current of the close circuit system produced from the PV module and that of delivered to the exhaust fan, respectively, for every two hours from 8.00 am to 6.00 pm for three consecutive days. Both Tests 2 and 3 are conducted on the same days (May 1–3, 2017). The ambient temperature and irradiance data for Test 3 are collected together with Test 2. And for Test 4, the data for the ambient temperature, voltage and current of the close circuit system produced from the PV module, voltage and current of the close circuit system delivered to the exhaust fan, and irradiance are collected for every two hours from 8.00 am to 6.00 pm for three consecutive days (May 4–6, 2017). All the collected data are tabulated, analyzed, and discussed in detail.

Several measuring devices and tools are used in this research to assist the data collection process. An anemometer was used to measure the airflow rate of the exhaust fan and the temperature readings for all the tests. A clamp meter was used to measure the voltage and current values needed in Test 3 and Test 4.

Figure 13 shows the weather station situated next to the research location. The irradiance data needed for the all the tests are collected through the weather station.

The collected data from all the tests are analyzed. The analysis of data consists of two parts, analysis of Test 1 and Test 2 for ventilation process and analysis of Test 3 and Test 4 for PV efficiency. The temperatures' data from Test 1 and Test 2 are analyzed. Graphs of the temperature data against time are plotted for both tests using Microsoft Excel. The differences in the ventilation process due to natural ventilation and solar-powered ventilation are analyzed and discussed. Additionally, the graph of irradiance against time is plotted to observe the variation of peak sun hour and weather condition throughout the testing days. The data from Test 3 and Test 4 are analyzed and the efficiency of the PV module is calculated. The efficiency of the PV module is calculated by:

$$\eta_{PV} = \frac{\text{Output power}}{\text{Irradiance} \times \text{Area of PV Module}} \times 100\% \qquad (2)$$

where the output power is the power produced from the PV module, irradiance is the radiant flux (power) received per unit area by the surface of the PV module from the Sun, and area of PV module is total surface area of the PV module which is 0.098 m^2. Graphs of the efficiency of PV module against the ambient temperature for different ranges of solar irradiance are plotted using Microsoft Excel. The variation of PV

Fig. 13 Weather station

efficiency with ambient temperature is observed and the efficiency of the PV module when there is cooling effect and that of when there is no cooling effect from the ventilation system is analyzed and discussed.

Findings

Ventilation Process

Figure 14 shows the indoor temperature for natural ventilation and solar-powered ventilation. The indoor temperature of the research location is lower during morning and is gradually increases and eventually higher than the ambient temperature as the day goes on until the late evening. This trend in the temperatures is observed for both natural ventilation and solar-powered ventilation as shown in Fig. 14. For natural ventilation, the results show that the indoor temperature starts to rise and is higher after about 9.00 am and continue to remain higher until the late evening for

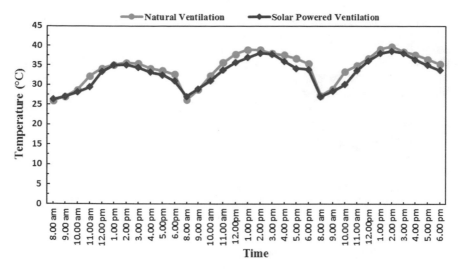

Fig. 14 Indoor temperature for natural and solar-powered ventilation

all the three days studied. Whereas for solar-powered ventilation, the results show similar trend as that of the natural ventilation but the indoor temperature is higher after about 10.00 am for all the three days. These results show that the solar-powered ventilation has kept the indoor temperature of the research location as low as possible at the beginning of the day compared to the natural ventilation. This can help reduce the cooling load of the research location and can ensure the late initialization of any air-conditioning system operation. The power consumption of the air conditioning system can be minimized as well as it does not have to cool down a very high indoor temperature in the beginning of its operation.

Figure 15 shows the temperature difference between the indoor temperature and ambient temperature for natural ventilation and solar-powered ventilation. It is noticed that the temperature differences between the indoor temperature and the ambient temperature are greater when the research location is undergoing natural ventilation. While when the research location is ventilated through the solar-powered ventilation, the temperature differences between the indoor temperature and the ambient temperature are smaller than that of natural ventilation. The range of temperature difference between the indoor temperature and the ambient temperature of the research location due to natural ventilation is 0.1–3.3 °C and that of due to solar-powered ventilation is 0.1–0.4 °C. It can be said that a more efficient solar-powered ventilation system has helped reduce the increased temperature difference between the indoor temperature and the ambient temperature during the hot days and kept it as low as possible by a good air circulation, improved air exchanging process, and a better ventilation process.

Thus, by having a solar-powered ventilation system instead of the natural ventilation, the indoor temperature of the research location can be reduced significantly, the

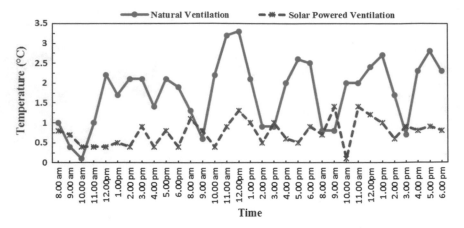

Fig. 15 Temperature difference between indoor temperature and ambient temperature for natural ventilation and solar-powered ventilation

cooling load and power consumption of any air-conditioning system in the research location can be minimized, and a better thermal comfort can be achieved.

Figure 16 shows the solar irradiance throughout the days of experiment of natural ventilation and solar-powered ventilation with and without cooling of the PV module. Almost same levels of irradiance and sunny weather were observed throughout the days of experiments. The solar insolation recorded over the period of three days for the experiment of natural ventilation was 5.1 kWh/m² and that of for the experiment of solar-powered ventilation with and without PV module cooling were 5.0 and 5.4 kWh/m² respectively. While, the maximum solar irradiance observed for the natural ventilation experiment was 750.26 W/m² and that of the solar-powered ventilation with and without PV module cooling experiments were 766.21 and 775.44 kWh/m² respectively.

Photovoltaic Efficiency

The cooling effect for the PV module is provided by the air ventilated out from the research location through the ventilation system designed in this research. It is an alternative passive cooling system for the PV module in which the ventilated air is directly channeled from the ventilation system to the back surface of the PV module. The presence of the air circulation just beneath the surface of the PV module will cool down and reduce the temperature of the PV module, which in turn will increase and maintain the efficiency of the PV module that are decreasing due to increasing temperature of the PV module during a hot day. Figures 17, 18, and 19 show the efficiency of PV module with and without cooling effect at different ranges of solar irradiance of 0–400, 400–700, and 700–1000 W/m², respectively. The efficiency of

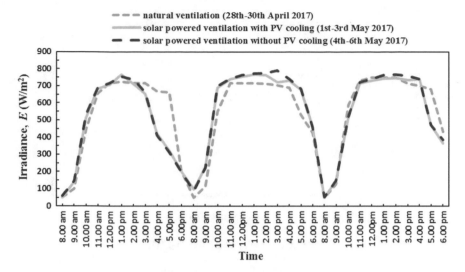

Fig. 16 Solar irradiance throughout the days of experiments

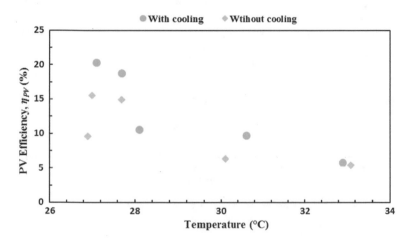

Fig. 17 Efficiency of PV module for the irradiance of 0–400 W/m^2

the PV module is highly depended on the irradiance, where a higher irradiance on the PV module increases its temperature and reduces its efficiency. The solar irradiance for the experiments of PV efficiency with and without cooling effect was shown in Fig. 18. The irradiance measured is used to calculate the efficiency of the PV module for both experiments as per Eq. (2) and the efficiency of the PV module is calculated by:

$$\eta_{PV} = \left[(\text{Output power})/(\text{Irradiance} \times \text{Area of PV Module})\right] \times 100\%$$
$$= \left[(4.7 \times 0.48 \text{ W})/(512.47 \text{ W/m}^2 \times 0.098 \text{ m}^2)\right] \times 100\% = 4.5\% \quad (3)$$

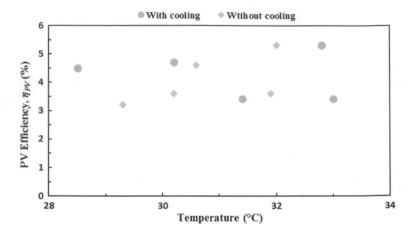

Fig. 18 Efficiency of PV module for the irradiance of 400–700 W/m^2

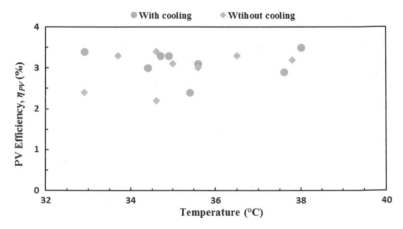

Fig. 19 Efficiency of PV module for the irradiance of 700–1000 W/m^2

Figure 20 shows the average efficiency of PV module at different ranges of solar irradiance for the experiments of PV efficiency with and without cooling effect. From Fig. 20, it is observed that the efficiency of the PV module decreases as the range of solar irradiance increases. It is seen in both experiments of PV efficiency with and without cooling effect. The highest average PV efficiency recorded for the experiment of PV efficiency with and without cooling effect is 13 and 10.3%, respectively, which is at lower irradiation range of 0–400 W/m^2, while the lowest average PV efficiency recorded for the experiment of PV efficiency with and without cooling effect is 3.1 and 3.0%, respectively, at higher irradiance range of 700–1000 W/m^2.

It is also observed from Fig. 20 that the average efficiency of the PV module with cooling effect is higher than that of without cooling effect for all the ranges of solar irradiance. However, it is noticed that better cooling effect for the PV module was

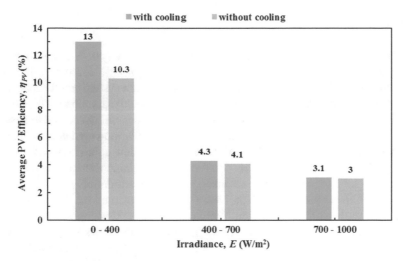

Fig. 20 Average PV efficiency at different ranges of solar irradiance

observed at lower irradiance of 0–400 W/m² as compared to irradiance ranges, and the effect decreases as the range of solar irradiance increases. At irradiance range of 0–400 W/m², the average efficiency of PV module increases by 26.2% when there is cooling effect on the PV module from the outflow of the ventilation system. It increases by only 4.9% at irradiance range of 400–700 W/m² and followed by an increment of only 3.3% at irradiance range of 700–1000 W/m². Besides, the average efficiency of the PV module recorded for the period of three days when there is cooling effect is about 6.2% and that of when there is no cooling effect is about 5.3%. This shows that the efficiency of the PV module increases by 17% if there is airflow (cooling effect) on the PV module from the outflow of the ventilation system developed in this research work. Therefore, this research finds out that the efficiency of the PV module decreases if the temperature of the PV module increases due to the higher solar irradiance on the solar panel.

Nonetheless, there are some contradictions over the data observed in Figs. 17, 18, and 19 where at a certain temperature range, the PV module gives two distinct efficiency values for the same temperature range. For instance, the efficiency of the PV module when the temperature reading is about 33 °C at the afternoon is about 3.4% and that of in the evening is about 5.4%. This is due to the reasons of different intensities of the irradiance throughout the day from morning to evening plus other environmental factors like wind speed, humidity.

In short, the analysis of this research sums up the performance, working condition, and functionality of the solar-powered ventilation system designed and developed in this research work. It is apparent that the developed solar-powered ventilation system can help enhance the ventilation process in a more efficient and effective way but lacks in its performance in cooling down the PV module.

Summary

An efficient and low-cost solar-powered attic ventilation system is designed and developed to facilitate the ventilation process of a hot attic space of a building. This efficient and effective solar-powered attic ventilation system could enhance the ventilation process of a building and at the same time could maintain the efficiency of the solar panel for a better performance, consequently benefiting each other systems and their performances simultaneously. Besides, the solar-powered attic ventilation system designed and developed in this research has a low capital cost and free of running cost if compared to the other commercially available ventilation system like the turbine ventilator or any other mechanical assisted ventilation.

Moreover, the experimental testing and the results obtained are the evidence for the performance, working condition, functionality of the developed solar-powered attic ventilation system. The results reveal that the developed system can reduce the attic temperature by 2.9 °C and capable of maintaining the temperature differences between the ambient temperature and the indoor temperature of the attic space in the range of 0.1–0.4 °C. Additionally, it can increase the PV efficiency by 17%, provided there is cooling effect from the outflow of the ventilation system beneath the surface of the PV module.

In short, the solar-powered attic ventilation system designed and developed in this research is economically low cost, able to reduce the hot attic temperature, lesser the cooling load and the power consumption of air-conditioning system, requires minimum maintenance, and environment friendly.

References

1. K.M. Al-Obaidi, M. Ismail, A.M.A. Rahman, A review of the potential of attic ventilation by passive and active turbine ventilators in tropical Malaysia. Sustainable Cities Soc. **10**, 232–240 (2014)
2. F. Basrawi, H. Ibrahim, M. Taib, G. Lee, Optimum thickness of wall insulations and their thermal performance for buildings in malaysian climate. Int. J. Automotive Mech. Eng. **8**, 1207 (2013)
3. R. Saidur, H. Masjuki, M. Jamaluddin, S. Ahmed, Energy and associated greenhouse gas emissions from household appliances in Malaysia. Energy Policy **35**, 1648–1657 (2007)
4. A. Al Yacouby, M.F. Khamidi, M.F. Nuruddin, A. Idrus, S.A. Farhan, A.E. Razali, A review on thermal performance of roofing materials in Malaysia, in *Sustainable Building and Infrastructure Systems: Our Future Today*, (2011), p. 351
5. K.M. Al-Obaidi, M.A. Ismail, A.M.A. Rahman, Effective use of hybrid turbine ventilator to improve thermal performance in Malaysian tropical houses. Building Serv. Eng. Res. Technol. **37**, 755–768 (2016)
6. American Society of Heating, Refrigerating and air-conditioning engineers (ASHRAE). *ASHRAE Handbook: HVAC Applications*, (American Society of Heating, Refrigerating, and Air-Conditioning Engineers Inc, Atlanta, 1999)
7. M.H. Ali, K.M.M. Billah, M. Mashud, Construction and performance test of a solar powered attic ventilation system, in *International Conference on Mechanical Engineering and Renewable Energy*, Chittagong, Bangladesh, Paper No. ICMERE2013-PI-069 (2013)

8. S. Ahmed, S.A. Rahman, and A. Zain-Ahmed, The ventilation performance of a solar-powered attic fan in Malaysian climate, in *Proceedings of the Conference on Sustainable Building South East Asia*, (2005), pp. 470–476
9. S. Dubey, J.N. Sarvaiya, B. Seshadri, Temperature dependent photovoltaic (PV) efficiency and its effect on PV production in the world–a review. Energy Procedia **33**, 311–321 (2013)
10. C. Misiopecki, A. Gustavsen, B. Time, Cooling of PV panels by natural convection, ZEB project report 6; SINTEF Byggforsk (2012)

Performance of Hydrogen Direct Injection Engine

Abdul Rashid Abdul Aziz, Muhammad Adlan Abdullah, Firmansyah and Ezrann Zharif Zainal Abidin

Hydrogen's high flammability, low ignition energy, clean burning, and high flame speed are attractive advantages of using internal combustion engines to attain high performance and efficiency as well as achieving "real" clean emission with only water and carbon dioxide as exhaust. To overcome volumetric efficiency loss, high NO_x emissions and abnormal combustion, the influence of air–fuel ratio is studied. Optimization of the air–fuel ratio effect and understanding of the combustion processes are actively being studied in the automotive industry and research institutions.

Introduction

Hydrogen economy is the most accepted and talked about hypothetical energy economy in the world today. Efforts to realize the hydrogen economy have led to research initiatives across the globe [1]. The hydrogen production from renewable resources supporting this initiative by introducing various production methods such as biomass gasification, biological production, biomass pyrolysis, and supercritical water hydrogen production [2]. Technology development of these production methods has reached commercialization stage with production efficiency of up to 70%, and thus improving the availability of hydrogen in the market and lowering the production costs while utilizing waste materials.

In the use of hydrogen for transportation fuel, there is a widely accepted argument that direct combustion of hydrogen in internal combustion engine could serve as a pathway to hydrogen economy before fuel cells technologies mature and become

A. R. Abdul Aziz (✉) · M. Adlan Abdullah · Firmansyah · E. Z. Z. Abidin
Universiti Teknologi PETRONAS, Perak, Malaysia
e-mail: rashid@utp.edu.my

M. Adlan Abdullah
PETRONAS Research Sdn. Bhd., Bangi, Malaysia

© The Author(s), under exclusive license to Springer Nature Singapore Pte Ltd. 2019
S. A. Sulaiman (ed.), *Sustainable Thermal Power Resources Through Future Engineering*, SpringerBriefs in Applied Sciences and Technology,
https://doi.org/10.1007/978-981-13-2968-5_2

cost effective [3]. Current automotive technologies and manufacturing processes can be used, and the experience of operating and handling gaseous fuels such as natural gas can prove beneficial. This would also facilitate proliferation of the hydrogen production and refueling infrastructures. As hydrogen has unique and sometimes contradicting characteristics in relation to its use as internal combustion fuel [4], it poses new challenges and opportunities. On the one hand, hydrogen's wide flammability range gives an opportunity for throttle-less operation (or power control by fuel quantity controlled means similar to diesel engines) with significant volumetric efficiency improvements potential. The high flame propagation speed gives rise to near constant volume combustion, yielding a potential thermal efficiency improvement. A further thermal efficiency improvement is also possible due to the inherent hydrogen's high auto-ignition temperature (a higher compression ratio engine is possible).

On the other hand, hydrogen's very low density and viscosity, low ignition energy, and low radiant heat push for new sets of thinking and approach to using it as a fuel for internal combustion engines. For example, while high compression ratio is desired considering the high octane of hydrogen, its low ignition energy may result in preignition due to glow ignition. In this case, optimized compression ratio or combustion chamber design may need to be determined. Even though the combustible limits of hydrogen are wide, its low density and viscosity may require optimization of injection parameters in order to reduce NO_x and the occurrence of preignition. Most of the previous works adopted lean burn strategies for hydrogen operation to avoid abnormal combustion. Even for direct injection engines, only compression ratios similar to gasoline engine were used.

The immediate development in direct hydrogen implementation as fuel is its utilization in a direct injection, spark ignition engine. In view of hydrogen favorable properties, it is used to overcome performance deficit of DI CNG engines, especially at low-end speeds (< 2500 rpm), whereby other gaseous fuels such as CNG have low combustion efficiency resulting in lower torque and power [4, 5].

In light of the hydrogen economy, hydrogen-fueled internal combustion engines are said to be the pathway or bridging technology before fuel cells technology becomes mature and widely available. This helps proliferate the hydrogen refueling infrastructure while gaining the much-needed experience and reputation. The experience of NGV industry can be proven valuable as the operation of the compressed gaseous fuel is not much different from hydrogen, especially in the aspects of refueling and storage [6]. With the increasing concern over global warming and the push for a hydrogen economy, the interest in hydrogen internal combustion engine (H2ICE) has recently been renewed. Some research groups in the automotive industry are looking into H2ICE, and some prototypes were launched.

Hydrogen internal combustion engine started as early as in the 1800 s when Francois Isaac de Rivaz of Switzerland used a mixture of hydrogen and oxygen as fuel [4]. The interest in hydrogen engine subsided as gasoline and diesel fuels, which have better versatility in terms of storage and energy density became widely available. With the advent of environmental and energy security issues coupled with the availability of the state-of-the-art engine controls, hydrogen internal combustion engines have recently regained much interest. This is compounded by the fact that

H2ICE is regarded as the transition technology before fuel cells technology is mature and ready for wide adoption.

The main driving force for H2ICE is the ability to operate the engine at significantly leaner air–fuel ratio so that higher efficiency in comparison to gasoline [7–9] can be achieved. This is due to the low lean limit and ignition energy of hydrogen that allows the engine to be run with stable combustion with little or no throttling. As such, most of the research works involving hydrogen internal combustion engines were aimed at increasing the lean limit of the spark ignition engines [10–12].

Hydrogen has stoichiometric heating value that is 17% higher than that of natural gas. Thus, it follows that for hydrogen, higher power output is expected for the same engine capacity. However, the specific output is generally low since hydrogen at stoichiometric occupies 30% volume in comparison to only 10% volume for natural gas. For comparison, gasoline occupies only 1.76% volume [14]. This is true in the case of carburetor or port fuel injection. For direct injection, the intake charge can be maximized, and hence, the full load performance can be increased [15].

Hydrogen's very wide flammability limit allows the engine to operate at ultra-lean combustion leading to high efficiency and low NO_x emissions. The engine can be run much, if not all the time, unthrottled. Its high octane number can be exploited to increase efficiency. Hydrogen combustion has a high flame velocity that leads to almost constant volume combustion, and thus resulting in a higher thermodynamic efficiency.

On the other hand, hydrogen's ignition energy is about one order of magnitude lower. Thus, it is susceptible to preignition from hot spots such as spark plug electrodes, combustion chamber deposits, oil contaminants, combustion in crevice volumes and residuals energy in ignition systems [16, 17]. This abnormal combustion could also lead to backfire as the combustion "flashback" to the intake system. As shown in Fig. 1 on the ignition energy versus equivalence ratio, operating the engine at close to stoichiometric air–fuel ratio increases the risk of the abnormal combustion [17–19]. Optimized fuel injection timing [20], water injection [21, 22], improved scavenging by variable valve timings, use of liquid hydrogen, ultra-lean combustion strategy, and direct injection [9, 20, 23] have been shown to be effective in reducing the preignition. Direct injection strategy also eliminates the possibility of backfire.

In addition, hydrogen combustion has higher flame temperature and lower quenching distance that leads to narrow thermal boundary layers. This results in more heat loss to the combustion chamber walls. Charge stratification strategies by employing direct injection technique was shown to be able to reduce the heat loss.

As hydrogen has no carbon content, its combustion results in zero fuel-derived carbon emissions. As expected, extremely low THC, CO, and CO_2 emissions are produced. The source of the minimal carbon emissions is from the engine lubricants. NO_x emissions can be an issue in hydrogen engine since it has high peak cylinder temperatures. It has been demonstrated that this can be effectively controlled by several means. Ultra-lean operation, EGR, water injection, use of liquid hydrogen, use of catalytic converter, and direct injection strategies are the various methods used to minimize the emission of NO_x.

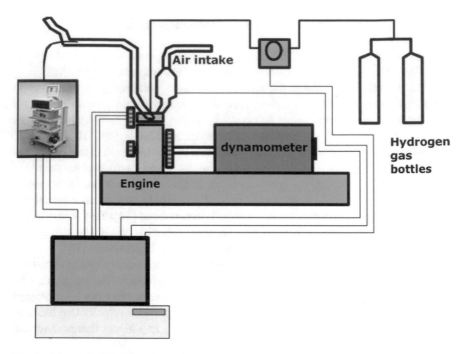

Fig. 1 Schematic of the experimental setup

In short, running an engine on hydrogen requires different strategies and design features than traditional fuels. Direct injection is generally preferred while the load control via mixture quality control means is desired for improved efficiency and fuel economy.

In an internal combustion engine, the energy introduced into the cylinder is closely related to the stoichiometric air–fuel ratio of the fuel in the air. The theoretical air–fuel ratio for hydrogen is 34 with volume percentage in the mixture of 29.6%.

For internal combustion engine, the energy introduced into the cylinder is the parameter that determines the mean effective pressure and power produced by the engine. For port/external mixture preparation, this depends on the mixture volume. In this case, since hydrogen occupies almost 30% of the mixture volume, the energy content is significantly diminished. In contrast, for direct injection, the energy introduced into the cylinder depends on the mass of air inducted into the engine.

The mixture calorific values of various fuels for external mixture preparation and for direct injection were shown by White et al. [13]. They reported that hydrogen has the largest difference in the mixture calorific value for direct injection and external mixture preparation as compared to other fuels such as methane. Thus, for hydrogen, direct injection is the solution for best power density. In the case of direct injection, hydrogen is also shown to have the highest mixture calorific value in comparison to other fuels. For example, hydrogen has 20% more mixture calorific value than natural gas (methane). Thus, it follows that for the same direct injection engine capacity, operating with hydrogen would result in 20% more torque and power than with natural gas.

The ignition energy required to start the combustion of hydrogen at various equivalence ratio is lower than that of natural gas. Therefore, it is expected that operating an engine with hydrogen would result in easier ignition in comparison with natural gas. This is also true at engine operating regions where air motion and mixing is poor.

The flame speed of hydrogen is approximately six times faster than natural gas, and it varies with different air–fuel ratio as found by Ilbas et al. [24]. Thus, it is expected that the combustion of hydrogen in an engine will be faster even when mixing is poor as in the case of low engine speed. These two arguments lead to the assumption that operating an engine with hydrogen would result in a more efficient combustion at lower speeds where natural gas is shown to have poor efficiency and lower performance.

Test Procedure and Equipment

Figure 1 shows a schematic of the experimental setup in this work. A single cylinder engine was coupled to a direct current dynamometer that allowed engine braking and motoring while the performance parameters were measured.

The Test Engine

The single cylinder engine system was set up based on a PROTON CamPro engine with modifications to its cylinder head to enable direct injection of gaseous fuel (Fig. 1). Shown in Fig. 2 is the cutoff view of the engine. Originally, the engine was designed for natural gas application. To demonstrate the practicality and easy adoption of hydrogen, no modification to the engine was made for this study. The specification of the engine is as given in Table 1. It was a four-stroke spark ignition engine with a compression ratio of 14:1 to take advantage of the high octane of natural gas. Since hydrogen auto-ignition temperature is higher than natural gas, this compression ratio was maintained.

A programmable ECU connected to a computer was used to control the engine. The engine parameters that could be controlled from the computer were the injection timing and duration, the spark timing, and throttle position. Real-time data were available from the engine ECU and could be viewed and recorded accordingly.

The original natural gas direct injector was used for the study without any modification, as shown in Table 2. Due to hydrogen's low density and viscosity, narrow-angle injector (30°) was chosen to allow maximum fuel spray penetration. The spray was executed with 18 bar injection pressure in air at atmospheric conditions. It shows that although the penetration of hydrogen was quite similar to natural gas, the distribution of hydrogen was wider. This suggests that the mixing of hydrogen was better than natural gas.

Fig. 2 Cutoff view of the
engine

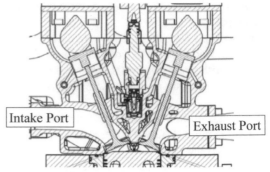

(a) Intake and exhaust port position

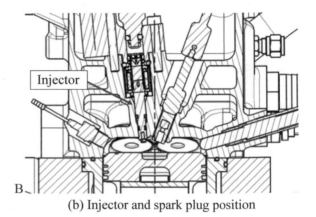

(b) Injector and spark plug position

The Dynamometer

Table 3 shows the specification of the direct current dynamometer. Being a DC
dynamometer, it has the capability to motor the engine. This was useful to obtain the
unfired cycle data for the cylinder pressure such that the combustion analysis can be
carried out. The data such as the torque, speed, engine oil temperature, coolant temperature, and intake air temperature were recorded manually from the dynamometer
panel.

The Fuel System

The fuel system used for the study is as shown in Fig. 3. The hydrogen was supplied
in gas bottles at a pressure of 200 bar. Pressure regulators were used to regulate the
pressure delivered to the engine. A Micro motion™ CMF010 ELITE Series fuel flow
meter was used to measure the fuel flow rate.

Table 1 Specifications of the single cylinder engine

Engine specifications	
Displacement volume	399.25 cm^3
Cylinder bore	76 mm
Cylinder stroke	88 mm
Compression ratio	14
Exhaust valve open	ATDC 10°
Exhaust valve closed	BBDC 45°
Inlet valve open	BTDC 12°
Inlet valve closed	ABDC 48°
Injection type	Direct injection, spray guided, central injector, 30° spray angle
ECU	Orbital Inc.
Positive crankcase ventilation	No
Injection pressure	18 bar

Table 2 Specifications of the fuel injector

Manufacturer	Synergist
Part number/designation	37-152 CNG
Nozzle spray angle	30°
Spring	33.0 N
Stroke	0.135 mm
Maximum operating pressure	2.0 MPa (high pressure)
Turn on time	1.05 ms
Turn off time	0.95 ms
Operating voltage	14.0 VDC
Coil resistance	1.3 O

Table 3 Dynamometer specifications

Make and model	David McClure DC30
Type	Direct current
Capacity	30 kW
Maximum speed	5000 rpm

The main limitation of the hydrogen implementation in ICE engine is its susceptibleness to abnormal combustion. Therefore, the discussion in this chapter starts with the abnormal combustion using hydrogen in relation to ignition timing and is followed by the effect of air–fuel ratio to the combustion performance of H2ICE.

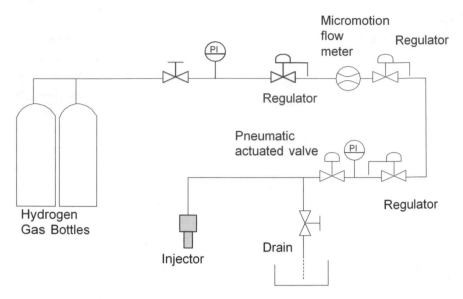

Fig. 3 Fuel supply system

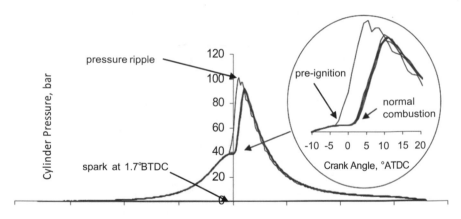

Fig. 4 Engine knocking caused by preignition

Abnormal Combustion in H2ICE

The abnormal combustion is highly affected by the auto-ignition temperature of hydrogen. The auto-ignition temperature of hydrogen is 574 °C, which is higher than that of natural gas (540 °C). However, abnormal combustion in the form of "knocking" was observed during the tests with hydrogen. When the engine was operated under certain conditions, pinging noise was detected accompanied by a sharp rise in the cylinder pressure and the corresponding pressure ripple, as implied in Fig. 4. These were the characteristics of engine knocking.

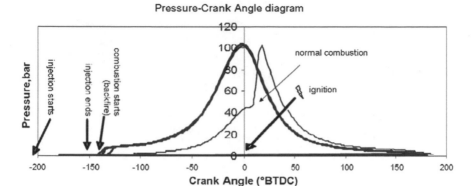

Fig. 5 Engine backfire due to early preignition

As shown in Fig. 4, combustion started prior to the spark. Thus, the knocking observed was thought to be due to preignition caused by hot spots in the engine. As the engine was cleaned prior to the test program, and the engine was inspected to be clean after the tests, the existence of hot spots from engine deposits was ruled out. Other possible sources were spark plug tip, sharp edges in the combustion chamber, or as a result of pyrolysis of engine oil.

At early injection timing and air–fuel ratio close to stoichiometric, there was a tendency for the preignition to start before the intake valve was closed, giving rise to the "backfire" condition. This is shown in Fig. 5. If the engine continues to operate with this abnormal combustion, damage to the engine intake systems will ensue.

These abnormal combustions led to the need to retard the ignition timing during the tests, especially at air–fuel ratios close to stoichiometric. Figure 6 shows the ignition timing for various speeds at different start of fuel injection (SOI). It is evident that for SOI of 300° BTDC, the ignition retards required to avoid preignition was less. This was due to the low volumetric efficiency and the subsequent lower combustion temperatures. The ignition retard was maximum at speed of 3000 rpm, i.e., the speed at which the volumetric efficiency was highest.

Figure 7 shows the variation of the ignition timing with respect to changes in air–fuel ratio. As the air–fuel ratio approached stoichiometrically, the combustion was more prone to preignition. In general, depending on the speed, there was approximately a 3–6° ignition retard needed for every 0.1 λ as it was approaching stoichiometric.

Figure 8 shows the corresponding cylinder pressures for different air–fuel ratios. It is evident that as a result of the retarded timing, the peak pressures were moving away from the typical optimum of about 15° ATDC as the air–fuel ratio was approaching stoichiometric. Thus, the maximum brake torque (MBT) timing was only possible on hydrogen operation at leaner mixtures. At close to stoichiometric, ignition timing was much retarded with a consequent penalty on the engine performance, as shown by the curve for lambda value of 1.16, which indicates that combustion starts very

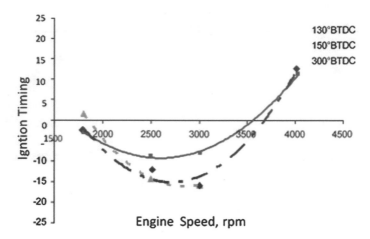

Fig. 6 Ignition timing for start of fuel injection of 130°, 150°, and 300° BTDC

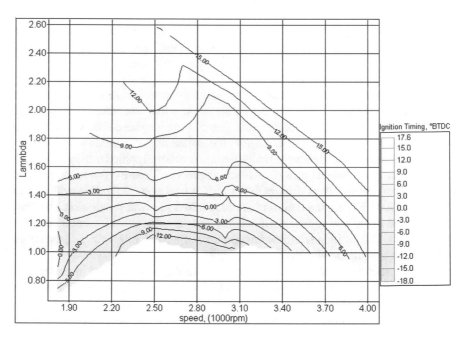

Fig. 7 Ignition timing map for stoichiometric air–fuel ratio of hydrogen

late in the expansion stroke. Thus, it is of interest to investigate whether operating at slightly leaner ratios with MBT timing would offset the performance deficit at stoichiometric.

It is important that for future efforts to adopt hydrogen for this engine, this abnormal combustion issue is addressed. It is even more important with higher engine

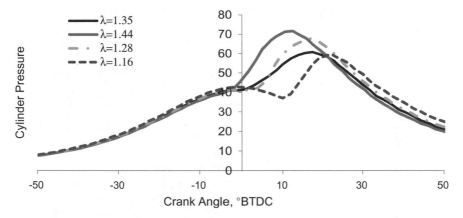

Fig. 8 Cylinder pressure showing the effect of retarding the ignition to avoid preignition at 3000 rpm

speeds as the engine is expected to be running hotter and the risk of melting the engine is higher. For all the tests conducted in this study, the best possible ignition timing without abnormal combustion was used throughout.

Effect of Air–Fuel Ratio to the H2ICE

Figure 9 shows the map of the engine torque across the various speed and the air–fuel ratios. It is worth noting that at leaner ratios ($\lambda > 1.2$), the ignition timing can be advanced and MBT timing can be achieved. However, as shown in the map, the increase in torque was not sufficient to offset the performance drop caused by the cleaning of the intake charge.

The torque map also shows the potential of controlling the power of the engine through mixture quality control method at low loads with the unthrottled operation. It was shown that the engine could be run at least to 50 or 60% load without throttle. The advantage of running the engine lean unthrottled is the increased thermal efficiency.

A map showing the indicated thermal efficiency across the engine speed and BMEP is depicted in Fig. 10. An indicated thermal efficiency of as high as 46% was achievable on this engine. Note, however, as the objective of the study is to improve the performance at full load, tests at higher air–fuel ratios were not conducted. Thus, the possibility of unthrottled operation at much lower loads was not assessed.

The emission of hydrocarbons is depicted in Fig. 11. In general, the emissions concentration is shown to decrease as the air–fuel ratio became leaner. As discussed in the previous section, the emission of hydrocarbon was probably originated from the lubricating oil and was related to quenching distance of the flame. As leaner ratios reduced the speed and lower the flame temperature, the quenching distance was higher. This explains the lower hydrocarbon emissions at leaner ratios.

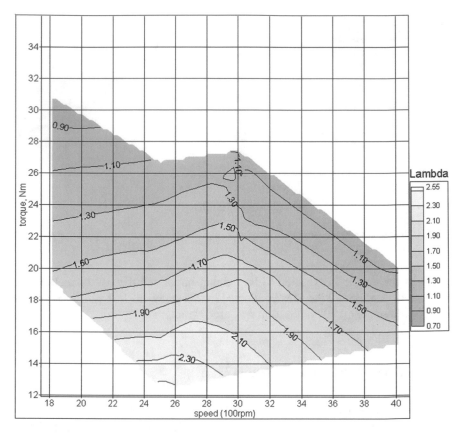

Fig. 9 Contour map of engine torque with respect to speed and air–fuel ratio (SOI = 130° BTDC except at 4000 rpm SOI = 160° BTDC)

As discussed in the previous section, the emission of CO is not significant at lean air–fuel ratio or close to stoichiometric. However, at the slightly rich air–fuel ratio, there were detectable CO emissions of up to 700 ppm. The possible explanation for this phenomenon is that the rich air–fuel ratio caused the flame to quench closer to the walls, resulting in oxidation of the engine oil film. This was more predominant at early injection timing, suggesting that the more homogenous is the mixture, the shorter would be the quenching distant.

NO$_x$ emission variation with changes in air–fuel ratio is also shown in Fig. 11. Overall, NO$_x$ emission was highest at lambda of around 1.1, at 3000–4000 ppm. At lambda ratios of more than 1.3, the NO$_x$ emission was reduced as leaner ratios resulted in lower temperatures, which is also illustrated in the figure showing the peak cylinder pressures. This agrees with the general understanding of NO$_x$ emission trend for air–fuel ratios of hydrogen. At stoichiometric air–fuel ratio, the NO$_x$ emission is similar to that at lambda of 1.3, suggesting that for low NO$_x$ operation, air–fuel ratio higher than lambda of 1.3 is desired.

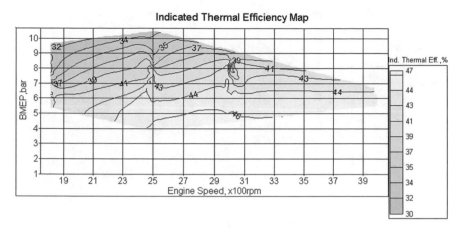

Fig. 10 Contour map of the engine's indicated thermal efficiency

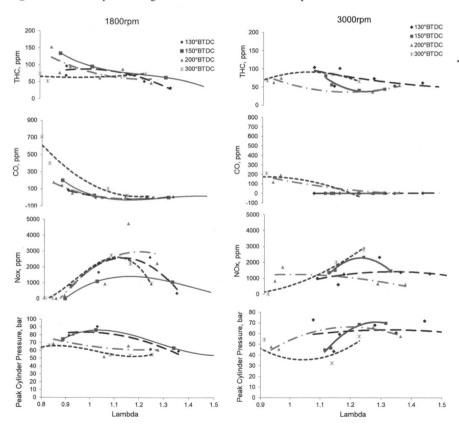

Fig. 11 Engine out emissions when using hydrogen at different air fuel ratios

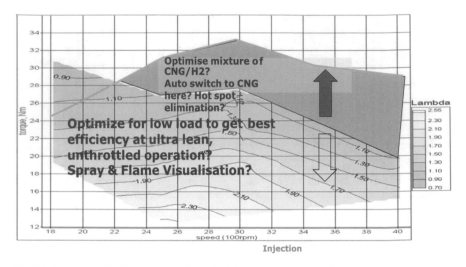

Fig. 12 Recommended future works for optimizing the hydrogen engine

Further optimization is required in order to fully operate the engine in hydrogen mode to cover the whole spectrum of engine operation. Figure 12 summarizes the recommended future works on this engine. In particular, the abnormal combustion needs to be eliminated. Cooler engine operations, reduction of hotspots in the combustion chamber, or running a mixture of hydrogen–natural gas are among the options available to be explored.

Hydrogen and natural gas could be simultaneously introduced into the engine at different ratios through port and direct injection, respectively. The effects on the combustion and flame propagation properties could be an important area of study to further optimize the performance and emissions. Optimum mixture ratios and timing of the injections may be determined to overcome the abnormal combustion while gaining efficiency and performance.

For low load applications, the unthrottled operation is recommended for thermal efficiency gain. In this mode, the injection duration is short, and opportunity exists for optimization in terms of injection parameters. The stratified charge could improve performance and allow very low loads such as idle to be run unthrottled. Flow and combustion visualization, as well as modeling, would help understand the processes so that optimization can be made.

Summary

Direct injection hydrogen enables operation with stoichiometric air–fuel ratio without abnormal combustion at low engine speed, i.e., below 2500 rpm. At higher engine speed, abnormal combustion limits the operation of the engine to leaner ratios or retarded ignition with significant performance penalty. Although MBT ignition timing can be achieved at slightly lean operation, the performance gained by the timing advance is not sufficient to overcome the loss due to the leaning effect. This is particularly obvious at low engine speed, 1800 rpm. Even though the efficiency and BSFC was minimum at lambda value of 1.2, the torque deficit was almost 5 Nm. Thus, for full load operation, stoichiometric operation is preferred at low engine speed.

The torque and power were relatively constant up to lambda value of 1.2 while the BSFC and efficiency were maximum at engine speed of 3000 rpm. Thus, for this speed, operation at lambda value of 1.2 was optimum.

Lean unthrottled operation of the engine leads to 46% indicated thermal efficiency. This mode of operation is suitable for lower load application such as during cruising. At this load and low engine speeds, the engine performance is less sensitive to the injection timing.

References

1. D.B. Osburn, Hydrogen & fuel cells fact sheet, 1–4. https://doi.org/10.1053/j.ajkd.2006.03.010
2. S.E. Hosseini, M.A. Wahid, Hydrogen production from renewable and sustainable energy resources: promising green energy carrier for clean development. Renew. Sustain. Energy Rev. **57**, 850–866 (2016). https://doi.org/10.1016/j.rser.2015.12.112
3. H. Council, *How Hydrogen Empowers the Energy Transition* (2017)
4. S. Verhelst, T. Wallner, Hydrogen-Fueled Internal Combustion Engines. Prog. Energy Combust. Sci. **35**, 490–527 (2009). https://doi.org/10.1016/j.pecs.2009.08.001
5. V. Chintala, K.A. Subramanian, A Comprehensive review on utilization of Hydrogen in a compression ignition engine under dual fuel mode. Renew. Sustain. Energy Rev. **70**, 472–491 (2017). https://doi.org/10.1016/j.rser.2016.11.247
6. R. Moliner, M.J. Lázaro, I. Suelves, Analysis of the strategies for bridging the gap towards the Hydrogen economy. Int. J. Hydrogen Energy **41**, 19500–19508 (2016). https://doi.org/10.1016/j.ijhydene.2016.06.202
7. F. Bisetti, J.Y. Chen, J.H. Chen, E.R. Hawkes, Differential diffusion effects during the ignition of a thermally stratified premixed Hydrogen-air mixture subject to turbulence. Proc. Combust Inst. **32**(I),1465–1472 (2009). https://doi.org/10.1016/j.proci.2008.09.001
8. F. Bisetti, J.-Y.Y.J.H. Chen, E.R. Hawkes, Probability density function treatment of turbulence/chemistry interactions during the ignition of a temperature-stratified mixture for application to HCCI engine modeling. Combust. Flame **155**, 571–584 (2008). https://doi.org/10.1016/j.combustflame.2008.05.018
9. A.R.A. Aziz, A. M-Adlan, M.F.A. Muthalib, Performance and emission comparison of a direct-injection (DI) internal combustion engine using Hydrogen and compressed natural gas as fuels. Int. Gas Res. Conf. Proc. **3**, 2224–2232 (2008)
10. G. Karim, Hydrogen as a spark ignition engine fuel. Int. J. Hydrogen Energy **28**, 569–577 (2003). https://doi.org/10.1016/S0360-3199(02)00150-7
11. F.Y. Hagos, A.R.A. Aziz, S.A. Sulaiman, Combustion characteristics of direct-injection spark-ignition engine fuelled with producer Gas. Energy Educ. Sci. Technol. Part A Energy Sci. Res. **31**, 1683–1698 (2013)
12. S.E.L. Mohammed, M.B. Baharom, A.R.A. Aziz, Analysis of engine characteristics and emissions fueled by in-situ mixing of small amount of Hydrogen in CNG. Int. J. Hydrogen Energy **36**, 4029–4037 (2011). https://doi.org/10.1016/j.ijhydene.2010.12.065

13. C.M. White, R.R. Steeper, A.E. Lutz, The hydrogen-fueled internal combustion engine: a technical review. Int. J. Hydrogen Energy **31**, 1292–1305 (2006). https://doi.org/10.1016/j.ijhydene.2005.12.001
14. R. Jorach, C. Enderle, R. Decker, Development of a low-NO_x truck Hydrogen engine with high specific power output. Int. J. Hydrogen Energy **22**, 423–427 (1997). https://doi.org/10.1016/S0360-3199(96)00083-3
15. L.M. Das, Hydrogen-Oxygen reaction mechanism and its implication to Hydrogen engine combustion. Int. J. Hydrogen Energy **21**, 703–715 (1996). https://doi.org/10.1016/0360-3199(95)00138-7
16. S. Szwaja, K.R. Bhandary, J.D. Naber, Comparisons of Hydrogen and gasoline combustion knock in a spark ignition engine. Int. J. Hydrogen Energy **32**, 5076–5087 (2007). https://doi.org/10.1016/j.ijhydene.2007.07.063
17. S. Pischinger, M. Günther, O. Budak, Abnormal combustion phenomena with different fuels in a spark ignition engine with direct fuel injection. Combust. Flame **175**, 123–137 (2017). https://doi.org/10.1016/j.combustflame.2016.09.010
18. Z. Wang, H. Liu, R.D. Reitz, Knocking combustion in spark-ignition engines. Prog. Energy Combust. Sci. **61**, 78–112 (2017). https://doi.org/10.1016/j.pecs.2017.03.004
19. G.T. Kalghatgi, Developments in internal combustion engines and implications for combustion science and future transport fuels. Proc. Combust. Inst. **35**, 101–115 (2014). https://doi.org/10.1016/j.proci.2014.10.002
20. A. Mohammadi, M. Shioji, Y. Nakai, W. Ishikura, E. Tabo, Performance and combustion characteristics of a direct injection SI Hydrogen engine. Int. J. Hydrogen Energy **32**, 296–304 (2007). https://doi.org/10.1016/j.ijhydene.2006.06.005
21. V. Subramanian, J.M. Mallikarjuna, A. Ramesh, Intake charge dilution effects on control of Nitric Oxide emission in a Hydrogen fueled SI engine. Int. J. Hydrogen Energy **32**, 2043–2056 (2007). https://doi.org/10.1016/j.ijhydene.2006.09.039
22. V. Subramanian, J.M. Mallikarjuna, A. Ramesh, Effect of water injection and spark timing on the Nitric Oxide emission and combustion parameters of a Hydrogen fuelled spark ignition engine. Int. J. Hydrogen Energy **32**, 1159–1173 (2007). https://doi.org/10.1016/j.ijhydene.2006.07.022
23. M. Talibi, P. Hellier, N. Ladommatos, The effect of varying EGR and intake air boost on Hydrogen-diesel co-combustion in CI engines. Int. J. Hydrogen Energy **42**, 6369–6383 (2016). https://doi.org/10.1016/j.ijhydene.2016.11.207
24. M. Ilbas, A.P. Crayford, I. Yilmaz, P.J. Bowen, N. Syred, Laminar-burning velocities of Hydrogen-air and Hydrogen-Methane-air mixtures: an experimental study. Int. J. Hydrogen Energy **31**, 1768–1779 (2006). https://doi.org/10.1016/j.ijhydene.2005.12.007

Waste Heat Recovery from a Gas Turbine: Organic Rankine Cycle

Mior Azman Meor Said and Muhammad Helmi Zin Zawawi

This chapter presents a study on optimization and parametrization of thermodynamic modelling of Organic Rankine Cycle (ORC) for waste heat recovery (WHR) of a gas turbine engine. The major aim of the study is to present a feasibility analysis in terms of thermodynamic performance for an integrated waste heat recovery system with a gas turbine system. The sections in this chapter provide detail out the findings at the preliminary stages.

In the current preliminary stage, comparison of performance analysis of ORC with that of Steam Rankine Cycle is executed. Thermodynamic modelling is carried out using Engineering Equation Solver (EES) software by using iteration methods to solve governing equations, fluid devices and heat transfer. The results are compared with previous studies from the literature in order to understand factors and limitations of ORC.

Background

Waste heat is ubiquitous in all industrial processes and machines and has potential to be utilized back in the main system. However, due to nature of the waste heat, which is low in temperature (less than 200 °C), conventional cycles, such as steam cycle, cannot be used to recover the heat [1, 2]. Nevertheless, ignoring the waste heat will be a big financial loss in the long run as it contributes to roughly half of the total energy of fuel burnt [3].

The Organic Rankine Cycle (ORC) uses the same feature as conventional Rankine Cycle except that it utilizes organic fluid as its working fluid. The lower boiling temperature of the organic fluid provides ORC the ability to harness low-grade waste

M. A. Meor Said (✉) · M. H. Z. Zawawi
Universiti Teknologi PETRONAS, Perak, Malaysia
e-mail: miorazman@utp.edu.my

© The Author(s), under exclusive license to Springer Nature Singapore Pte Ltd. 2019   37
S. A. Sulaiman (ed.), *Sustainable Thermal Power Resources Through Future Engineering*, SpringerBriefs in Applied Sciences and Technology,
https://doi.org/10.1007/978-981-13-2968-5_3

heat [4]. The ORC has been developed since the early 1980s and has been applied to power plants for geothermal and combined heat and power [5].

Since the working principle of ORC is the same as that of Rankine Cycle, the waste heat is converted into mechanical power, except with ORC; this regeneration process can be done more efficiently. Despite having the same working principle, the optimization of the ORC is a bit different from that of steam cycle, due to the limitations on the heat source temperature and the absence of constraint for vapour quality at the end of the expansion [5].

ORCs have never been in so much demand until now due to today's growing concern over the future depletion of fossil fuels and global environmental destruction. Most manufacturers and operators are turning their interest on low-grade energy recovery systems. The ORC's low operating temperature has made it possible for it to recover heat from various sources. ORC in waste heat recovery has helped to reduce thermal pollution tremendously while contributing in energy conservation. ORC systems are already being used in many countries in the world such as USA, Canada, Italy and Germany [6]. However, this technology is still considered new in Malaysia. This chapter focuses on waste heat recovery of a gas turbine engine to investigate the parameters and factors that govern the optimization of ORC. Once the parameters and factors have been determined, one can proceed to the next step in improving ORCs.

The current study highlights the optimization of ORCs by using thermodynamic models and simulates these models in designated software. The specific objectives of this project are:

a. To investigate ORC performance using thermodynamic modelling for waste heat recovery of a gas turbine engine.
b. To study the difference between thermal efficiency of Organic Rankine Cycle and that of Steam Rankine Cycle.
c. To investigate the effect of inlet temperature of turbine towards the performance of Organic Rankine Cycle.

Working Principle of ORC

The basic thermodynamic processes of an ORC are the same for Rankine Cycle. Rankine Cycle is an energy producing thermodynamic closed-loop cyclic system that has working fluid flows through four components: evaporator, condenser, pump and expander [7]. Energy or heat is absorbed into the system via the evaporator at constant pressure, while the energy is released from the system via condenser. During the heat absorption at the evaporator, phase change occurs where working fluid change phase from saturated liquid to saturated or superheated vapour, while in the condenser phase change, process occurs where vapour is condensed to saturated or under cooled water.

Fig. 1 Basic layout of
Organic Rankine Cycle

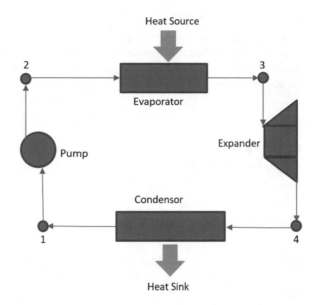

Expander is used to reduce the pressure of working fluid that exits the evaporator, before entering the condenser. In the expander, energy is transferred to mechanical work at constant entropy. Pump is used to increase the pressure of working fluid that exits the condenser, before entering the evaporator. These processes form the basic cycle of an ORC.

System Equations and Theoretical Analysis

In order to understand the working of ORC itself, it is essential to understand the working principles of each phase in ORC. Figure 1 shows the basic layout of Organic Rankine Cycle which involves the following processes: Compression (Nodes 1–2), Heat Absorption (Nodes 2–3), Expansion (Nodes 3–4) and Heat Release (Nodes 4–1).

Compression (Nodes 1–2)

A pump compresses the working fluid from saturated liquid state to high-pressured liquid state, before entering the evaporator. This process occurs at constant entropy. The power required by the pump can be estimated by:

$$\dot{W}_{1-2} = \dot{m}(h_2 - h_1) \tag{1}$$

where $\dot{W}_{1-2}$ is the power required by pump, $\dot{m}$ is the mass flow rate, h_1 is the enthalpy at pump inlet and h_1 is the enthalpy at pump outlet.

Energy is defined as the limit of useful work during a thermodynamic process that involves energy work transfer with a heat reservoir. The energy destruction rate of a pump is given by:

$$\dot{X}_{1-2} = T_0 \cdot (s_2 - s_1)\dot{m} \tag{2}$$

where $1-2$ is the energy destruction rate of a pump, T_0 is the ambient temperature in K, s_1 entropy at pump inlet and s_2 is the entropy at pump outlet.

Heat Absorption (Nodes 2–3)

At the evaporator, heat is absorbed by the working fluid in which the working fluid undergoes phase change from saturated liquid to saturated or superheated vapour. The process is assumed as isobaric, although there is a slight pressure drop in evaporator tubes. The heat absorption rate during this process can be described by:

$$Q_{2-3} = \dot{m}(h_3 - h_2) \tag{3}$$

where Q_{2-3} is the heat absorption rate at the evaporator, h_3 is the enthalpy of working fluid at the exit of refrigerator tube and h_2 is the enthalpy of working fluid at the inlet of refrigerator tube.

Energy destruction rate inside evaporator can be described by:

$$\dot{X}_{2-3} = \cdot T_0 \cdot \left[(s_3 - s_2) - \frac{(h_3 - h_2)}{T_M} \right] \cdot \dot{m} \tag{4}$$

where $\dot{X}_{2-3}$ is the rate of energy loss in the evaporator, while s_3 is the entropy of the vapour. T_M is the mean temperature along the evaporator.

Expansion (3–4)

The energy of the working fluid will be converted into useful mechanical work by expander or turbine in the expansion process. The process is assumed to be an isentropic process. The power output produced by turbine can be calculated by:

$$W_{3-4} = \dot{m}(h_3 - h_4) \tag{5}$$

where W_{3-4} is the power output of the turbine or expander, h_4 is the enthalpy of the working fluid at the outlet of the turbine.

The rate of energy loss in the turbine or expander can be shown by:

$$\dot{X}_{3-4} = T_0 \cdot (s_4 - s_3) \cdot \dot{m} \tag{6}$$

where $\dot{X}_{3-4}$ is the rate of energy loss in the expander or turbine while s_4 is the entropy of the working fluid at turbine or expander outlet.

Heat Release (4–1)

At the condenser, the working fluid releases heat via condensation process that change from saturated vapour phase to saturated or under cooled liquid phase. The process is assumed under constant pressure although there is pressure reduction along with the condenser tube due to friction against the wall. The heat release rate in the condenser can be shown by:

$$Q_{4-1} = \dot{m}(h_4 - h_1) \tag{7}$$

where Q_{4-1} is the heat release rate in the condenser.

In order to compute the rate of energy loss at the condenser, mean temperature along with the condenser tube can be introduced, as given by:

$$\dot{X}_{4-1} = \left[(s_1 - s_4) - \frac{(h_1 - h_4)}{T_M} \right] \dot{m} \cdot T_0 \tag{8}$$

where $\dot{X}_{4-1}$ is the rate of energy loss at the condenser.

Thermal efficiency of a thermal system described the effectiveness of the system to produce the desired output over the heat input. Thermal efficiency of an Organic Rankine Cycle is the ratio of the net output work of the system to the heat absorbed at the evaporator, given by:

$$\eta_{\text{eff}} = \frac{(W_{3-4} - W_{1-2})}{Q_{2-a}} \tag{9}$$

Organic Rankine Cycle Modelling

Commercial engineering software, Engineering Equation Solver (EES), is used to model the ORC and the Steam Rankine Cycle. The principle behind the EES computation is solving mass and energy governing equations at all specific nodes via numerical methods. Equations (1)–(9) in the previous section are also computed as part of the iterations.

The simulation is done by setting up EES environment parameters as shown in Table 1. In this simulation, efficiency of Organic Rankine Cycle is compared with

Table 1 List of EES
environment parameters

Parameters	Values
Pressure at location 1, P_1	3630 kPa
Temperature at location, T_1	Range from 60 to 130 °C
Turbine efficiency	0.7
Pump efficiency	0.7
Net power output	1000 kW
Temperature at location 2, T_2	−20 °C
Temperature at location 3, T_3	$(T_2 - 15)$ °C
Pressure at location 3, P_3	P_2
Pressure at location 4, P_4	P_1
T_{abs}	$(T + 273.15) \times 1.8$ Rankine

Steam Rankine Cycle. The net output work is also one of the gauges to determine which working fluid is better. The first simulation is done by using organic fluid, isobutane and varying the temperature for the boiler. Pressure of the boiler is set to be the critical pressure of isobutane as done by Nouman [6]. Since pressure and temperature exiting the boiler are equal to that of entering the turbine, varying these values for boiler directly alter the value for turbine. Iteration is carried out when it comes to the temperature that enters the turbine.

Once iteration is completed for simulation using isobutane, another simulation is carried out using steam in order to differentiate the efficiency between Organic Rankine Cycle and Steam Rankine Cycle. It is important to note that for second simulation, values of boiler pressure and temperature from the first simulation are used.

Results and Discussions

Before thermal efficiency and net output work are discussed, first and foremost, property graphs of isobutane and steam have been plotted in *T-s* diagram, *T-v* diagram and *P-v* diagram. The graphs plotted display the thermo-physical properties of a dry fluid for isobutane. In Figs. 2 and 3, isobutane clearly shows positive saturation vapour line (SVL), whereas steam shows negative slope for SVL. As stated by Nouman [6], positive slope in the T-s diagram is favorable for turbo machinery, as the working fluid leaves the expander as superheated vapour. This can reduce the risk of corrosion. Besides, the need to overheat the vapour before entering the expander can be eliminated, and a smaller and cheaper evaporator can be used.

Fig. 2 *T-s* diagram of
isobutane

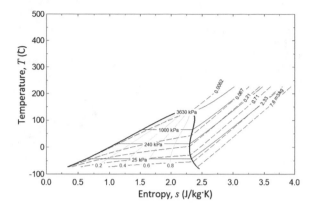

Fig. 3 *T-s* diagram of steam

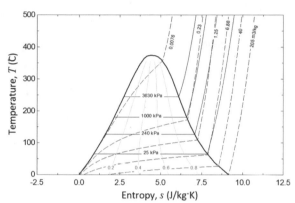

Besides, *T-s* diagram, graphs of *T-v* diagram and *P-v* diagram are also plotted. Figures 4, 5, 6 and 7 further validate the thermo-physical properties of isobutane as dry working fluid. Moreover, it draws the significant difference of thermo-physical properties of steam and that of isobutane in term of boiling point, critical pressure and temperature and others.

Next, the results obtained from the optimization of Organic Rankine Cycle are discussed. Values tabulated and graph plotted are compared with one of the previous studies done by Nouman [6].

Table 2 shows the value obtained from completing the two simulations. The value of T_1 shows the temperature of turbine. As explained in the previous chapter, the value is optimized in order to ensure the simulation which is carried out with no error, by referring to thermo-physical properties of isobutane. Nouman [6] stated that at critical pressure of 3630 kPa for isobutane, the critical temperature would be 134.7 °C. From the first simulation, any attempt to vary the temperature higher than critical temperature of 134.7 °C will only be met with error. Hence, first simulation is carried out within the critical temperature of isobutane. The second simulation with steam is carried out using the same values for T_1. The reason is to see the efficiency of steam under the same operating condition of isobutane.

Fig. 4 T-v diagram of
isobutane

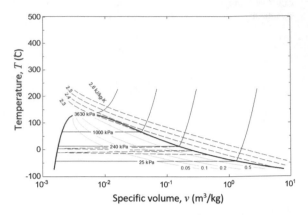

Fig. 5 T-v diagram of steam

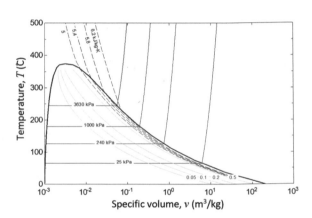

Fig. 6 P-v diagram of
isobutane

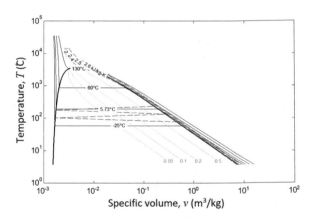

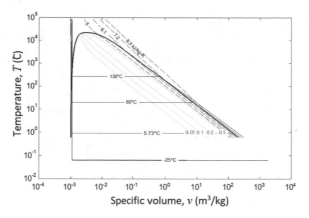

Fig. 7 P-v diagram of steam

Table 2 Values for efficiency and net output work

Run	P_1 (kPa)	T_1 (°C)	η_{th} isobutane	η_{th} steam	W_{net} isobutane (kW)	W_{net} steam (kW)
1	3630	60.00	0.06373	0.06692	13.92	43.57
2	3630	67.78	0.07334	0.07202	17.55	49.23
3	3630	75.56	0.08256	0.07719	21.51	55.28
4	3630	83.33	0.09146	0.08242	25.83	61.71
5	3630	91.11	0.10010	0.08767	30.52	68.50
6	3630	98.89	0.10850	0.09293	35.63	75.65
7	3630	106.70	0.11680	0.09818	41.21	83.14
8	3630	114.4	0.12520	0.10340	47.40	90.96
9	3630	122.2	0.13370	0.10860	54.47	99.11
10	3630	130.0	0.14340	0.11380	63.42	107.60

Since the difference in organic fluid with steam is the low heat source temperature for evaporation and recovering thermal energy from low-grade heat sources, tabulated value for thermal efficiency η_{th} and graph plotted have shown higher efficiency in isobutane compared with the value for steam. From the results, the efficiency of isobutane is increased within the range of 1–12%, as adhered by Li [8]. The percentage difference is shown in Table 3.

From the results, relationship between efficiency of isobutane and steam with temperature of turbine shows linear relationship. The increment is linear, and thermal efficiency of isobutane is higher than that of steam. However, at 60 °C, thermal efficiency using isobutane is lower than that of steam. The reason is because this temperature is not optimum working temperature for isobutane in comparison with steam.

Table 3 Percentage difference of efficiency

T_1 (°C)	η_{th} isobutane	η_{th} steam	Percentage difference (%)
60.00	0.06373	0.06692	−2.44
67.78	0.07334	0.07202	0.9
75.56	0.08256	0.07719	3.36
83.33	0.09146	0.08242	5.19
91.11	0.10010	0.08767	6.62
98.89	0.10850	0.09293	7.73
106.70	0.11680	0.09818	8.66
114.4	0.12520	0.10340	9.54
122.20	0.13370	0.10860	10.36
130.00	0.14340	0.11380	11.50

Table 4 Comparison of thermal efficiency with other work

T_1 (°C)	η_{th} isobutane	η_{th} isobutane [6]	Percentage difference (%)
60	0.06373	0.060	−3.01
67.78	0.07334	0.080	4.74
75.56	0.08256	0.090	4.31
83.33	0.09146	0.100	4.46
91.11	0.10010	0.110	4.71
98.89	0.10850	0.115	2.91
106.7	0.11680	0.122	2.18
114.4	0.12520	0.130	1.88
122.2	0.13370	0.135	0.48
130	0.14340	0.140	−1.20

The net output work shows the difference of mechanical work done between turbine and pump. As shown in Table 2, W_{net} for isobutane is lower than W_{net} for steam as the highest value of W_{net} for isobutane is 63.42 kW while for steam is 107.6 kW. This is expected since higher energy is needed to convert steam from one phase to another. Less work is required for isobutane as one of the advantages of organic fluid is its low boiling point.

Table 4 shows the percentage difference between values of thermal efficiency. One is from the simulation done and another is from research work done by Nouman [6]. Both parameters used in these simulations are similar. The value of percentage difference varies from 0.48 to 4.74%.

Summary

The selection of optimal working fluid for Organic Rankine Cycle is a difficult process considering that there are many available working fluids to choose from. These working fluids have different thermodynamic properties that are desirable to different type of set up for Organic Rankine Cycle.

Since this study is done within the scope of subcritical Organic Rankine Cycle, the evaporation and condensation pressures need to be lower than the critical pressure for isobutane for optimum condition. The fluid selection process is a trade-off between thermodynamic, environmental and safety properties. However, due to limitation of time, this study only focuses on isobutane and the comparison with steam. From this study, it is clear that the working fluid, namely organic fluid, isobutane gives higher value when it comes to thermal efficiency by 1–12% as compared to steam.

For future works, another factor that should be included in this study is the economic feasibility. It is important to find out whether Organic Rankine Cycle is feasible to be implemented in Malaysia.

Besides, we can also further our understanding by using different organic fluid that covers Transient Cycle and Supercritical Phase. This will be helpful in choosing the organic fluid that gives the best performance of Organic Rankine Cycle for waste heat recovery of a gas turbine engine. This study will be useful in assisting manufacturers in using the most optimum organic fluid for Organic Rankine Cycle.

References

1. EPA, Industrial waste heat recovery and the potential for emissions reduction, **1** (1984)
2. S. Quoilin, R. Aumann, A. Grill, A. Schuster, V. Lemort, H. Spliethoff, Dynamic modeling and optimal control strategy of waste heat recovery organic Rankine cycles. Appl. Energy **88**(6), 2183–2190 (2011)
3. H. Kim, W. Kim, S. Kim, Applicability of scroll expander and compressor to an external power engine: Conceptual design and performance analysis. Int. J. Energy Res. **36**(3), 385–396 (2011)
4. J.P. Roy, M.K. Mishra, A. Misra, Parametric optimization and performance analysis of a waste heat recovery system using organic Rankine cycle. Energy **35**(12), 5049–5062 (2010)
5. A.B. Little, S. Garimella, Comparative assessment of alternative cycles for waste heat recovery and upgrade. Energy **36**(7), 4492–4504 (2011)
6. J. Nouman, *Comparative Studies and Analyses of Working Fluids for Organic Rankine Cycles-ORC* (2012)
7. H. Chen, D.Y. Goswami, E.K. Stefanakos, A review of thermodynamic cycles and working fluids for the conversion of low-grade heat. Renew. Sustain. Energy Rev. **14**(9), 3059–3067 (2010)
8. Y. Li, *Analysis of low Temperature Organic Rankine Cycles for Solar Applications* (2013)
9. T. Yamamoto, T. Furuhata, N. Arai, K. Mori, Design and testing of the organic Rankine cycle. Energy **26**(3), 239–251 (2001)

Solar Thermal Energy Trapping Mechanisms in Practice

Syed Ihtsham-ul-Haq Gilani

Sun is the only source of energy to the earth on which the whole global activities are dependent. Solar energy is the starting point for natural, chemical and biological processes in this world in the early days and it is the most environmental-friendly form of energy. It can be used in many ways and it is suitable for many sorts of solar/chemical systems.

Solar energy is beneficial to mankind in two ways, i.e., in terms of light and heat. Mankind has a long history of using solar energy. Its only during the past century, mankind developed the equipment that could convert the solar energy into useful forms of energy. One of the main equipment that absorbs the heat from the sun is called solar collector. It uses one of the most important techniques that could be employed to increase the utilization of solar radiation in solar thermal heating technologies. The potential of solar irradiance is approximately 1366 W/m^2 at the radiation wavelength range from 0.3 to 3 μm, which is the visible range of radiation. Other wavelengths also carry energy, but the visible range is the most important in solar applications. Detailed descriptions of solar trapping mechanism, including its categories, applications, component of different categories, operational concept and design, in addition to the historical view and current status are discussed in this chapter.

Solar Trapping Mechanism/Equipment

The basic concept of solar energy trapping mechanism is based in using of glazing envelope to save the harvested heat on a heat transfer fluid (HTF) that flows inside the solar absorber tubes. High optical transmittance of glass has made it the most

S. I. Gilani (✉)
Universiti Teknologi PETRONAS, Perak, Malaysia
e-mail: syedihtsham@utp.edu.my

© The Author(s), under exclusive license to Springer Nature Singapore Pte Ltd. 2019   49
S. A. Sulaiman (ed.), *Sustainable Thermal Power Resources Through Future Engineering*, SpringerBriefs in Applied Sciences and Technology,
https://doi.org/10.1007/978-981-13-2968-5_4

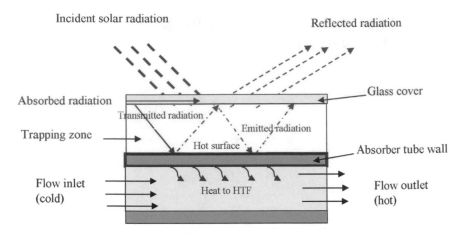

Fig. 1 Cross-sectional view of trapping zone and heat transfer in a solar collector

wanted candidate for solar glazing envelope. Glass has a unique characteristic that it allows shortwave radiations coming from the sun to pass through it, but it does not allow the long-wave radiation from hot bodies to pass through it. It means the major passage of heat energy is from sun to the absorber and less energy is lost in the form of long-wave radiations. At the same time, the glass has low thermal conductivity, which is protecting the thermal gain to be lost to the surrounding space.

From this state, high-temperature trapping zone is created in the space between the absorber tube and the glazing envelop as shown in Fig. 1; that is why, the most of the energy entering the glazing zone is captured by the absorber and transferred to the heat transfer fluid. If the losses from the absorber are controlled, then the absorber temperature becomes directly dependent on the solar radiation intensity.

General Classifications of Solar Thermal Systems

Shown in Fig. 2 is the comprehensive classification of solar trapping mechanism. Thermal collectors are divided into many categories depending upon their use as well as their output. They may be classified on the basis of their reflection mechanism. Solar collectors are divided into three main categories on the basis of their output temperatures, i.e., low temperature (>85 °C), medium temperature (>150 °C), and high temperature (1000 °C).

For all the three categories, the equipment is also specified separately. For low temperature, flat-plate collectors are quite popular, for medium temperatures, evacuated tube collectors can play an important role, and for high-temperature applications, only concentrated solar power technology would be serving the purpose. Flat-plate collector can be further divided into liquid base or air base, depending upon the

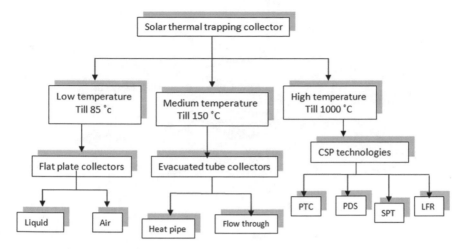

Fig. 2 Classification of solar thermal trapping collectors

HTF, whereas evacuated tubes technic is used in collectors having flow through or heat pipe technology. Generally, the use of CSP technology is executed by parabolic troughs, parabolic dish systems, solar power tower system, or linear Fresnel receivers, discussed in later sections.

Categories of Solar Trapping Mechanism

Currently, solar trapping mechanism is utilized in different applications such as solar cooker, water heater, and steam generators. Solar heat trapping and using for later hours or for daily uses is an old technology used by the mankind and it can be categorized according to operational temperature limits to low-, medium, and high-temperature applications as described in the following sections.

Low-Temperature Solar Trapping Collectors

The low-temperature solar trapping collectors are designed for applications that require energy delivered at temperatures up to 85 °C. Solar radiations can be divided into two types, i.e., beam radiation and diffuse radiation, and both the beam and the diffuse solar radiation are utilized in this category of solar application.

Many design approaches are suggested and applied in this category including simple and modified flat-plate collectors. The design is mainly based on direct practicing of the simple trapping concept in which the collector design is only composed of absorber tubes, glazing cover, and supporting collector box. Two basic types are

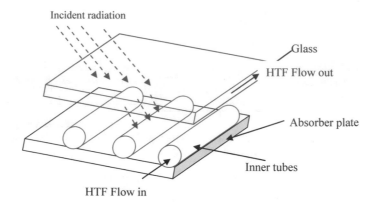

Fig. 3 Schematic of a flat-plate solar collector

currently commercialized on this category based on the HTF type, i.e., liquid or air types [1]. The air-type solar collectors are used for relatively higher temperature ranges.

A number of researchers used various techniques to improve the performance of these types of collectors. Few such techniques are the use of finned absorber, control of the flow regime, use of extra absorber plate to improve tube absorbance, control of the relative position between absorber tubes and plate, i.e., center, top, or under position. Some researchers used different paint on the absorber plate or absorber tubes to increase the absorptivity of the material, resulting in more absorption and high-temperature gains. A schematic of a flat-plate solar collector is shown in Fig. 3, where incident solar radiation is transmitting from the glass and absorbed by the absorber tubes and plate; thus, heat absorbed is added to the HTF flowing through the absorber tubes, which is termed as heat gain of the solar collector.

Medium Temperature Solar Trapping Collectors

The collectors in this category are designed for applications that require energy delivered at an elevated temperature of 150 °C. A modified trapping mechanism is used to increase the utilization of the received beam and diffuse solar radiation. These modifications are based on the use of both evacuation and simple concentration techniques, as compound parabolic concentrator.

Since vacuum is the best insulator and heat cannot pass through it, vacuum is used to improve the absorber insulation to avoid heat losses. Since absorber (metal) tube is surrounded by the glass tube and the air between absorber and glass tubes is removed, the absorber tube can only lose heat through the heat transfer fluid flowing through it and no heat energy is lost to the environment through convection

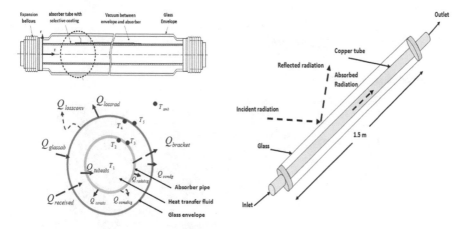

Fig. 4 Schematic view (left) and diagram (right) of the flow through type evacuated tube configuration

or radiation to the surrounding. Many design approaches are suggested and applied in this category including heat pipe with different shapes and pass through collectors. A cross-sectional view of the system is shown in Fig. 4.

High-Temperature Solar Trapping Collectors

To create the high temperature, in the range of 250 °C and above, the solar radiation needs to be added or multiplied; otherwise, they have the intensity of 1366 W/m². To increase the solar intensity, reflectors are used to reflect the sunlight from a larger surface area to an absorber having less surface area. In this way, the ratio of the areas becomes the multiplication factor to the solar energy, thus producing high temperatures. These types of collectors are known as concentrating solar power (CSP) units. They are used in applications of heat for industry and steam generation. In high-temperature solar trapping collectors, energy is delivered at temperatures up to 1000 °C above ambient temperature. Concentrators are used to elevate the received amount of direct solar radiation to degrees more than 1000 W/m². Usually, advanced solar tracking mechanisms are used as auxiliary with high-precision control system.

Different design approaches are suggested and applied in this category including parabolic trough collector (PTC), parabolic dish collector (PDC), linear fresnal reflector (LFR), and solar power tower (SPT). In this category, collector design is complicated and many subsystem components are shared to execute the heating process. One of the major drawbacks in this type of systems is a handsome amount of heat which is lost to the environment, because the operating temperatures are quite high, thus losing heat through convection as well as radiation modes of heat transfer.

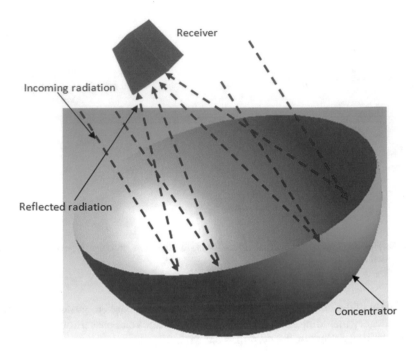

Fig. 5 Schematic of the parabolic dish collector

Differences between various CSP technologies are mainly in the concentrator and receiver element's configurations and sizes, which consequently affect the applied tracking mode, capability, elevated power, receiver design, and operating temperature, etc. In the PTC, PDS, and LFR technologies, the unit is categorized from small to medium size and capability, the receiver is an integrated part of the unit, and the plant is a field of multi-units generating the required power, where the solar power tower SPT is a unified large-scale system. Schematics of PDS, PTC, FLR, and SPT systems design and how radiation received and concentrated to the HTF are shown in Figs. 5, 6, 7, and 8. Parabolic dish system is concentrating the solar radiation on a point as they solar power is doing the same, whereas the parabolic trough collector system and Fresnel lens system concentrate the solar radiation to a line, where a receiver tube is installed.

Concentration ratio of PTC and dish cannot be extended beyond a certain value because of the system stability problems. However, in the case of the solar power tower and Fresnel lens system, the solar concentration ratio can be increased to the resources limit, and therefore, they can provide much higher temperatures.

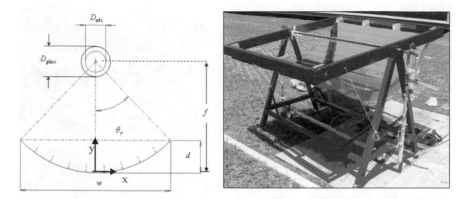

Fig. 6 Schematic view of parabolic trough collector (left) and a parabolic trough photograph at UTP

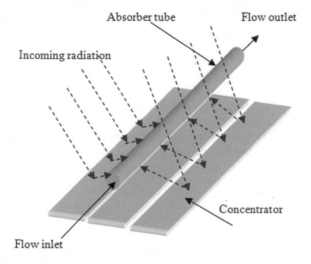

Fig. 7 Linear fresnal reflector

Thermal Analysis of the Solar Collector Performance

A detailed thermal analysis of the collector design parameters and their effect on the collector performance could be understood from a mathematical model, which is published by Duffie [2]. The model states that the collector's useful heat gain can be attained by:

$$Q_{useful} = A_c \left[S - U_L \left(T_{pm} - T_a \right) \right] \tag{1}$$

Fig. 8 Solar power tower model installed in UTP

where Q_{useful} is representing the useful heat gain, A_c is the collector area, S is the valuable radiation received to the absorber, U_L is the heat transfer factor that depends upon the absorber tube material and collector geometry, and $(T_{pm} - T_a)$ is the temperature differences between absorber surface temperature and ambient temperature.

On the other side and in term of HTF thermal analysis, the same amount of collector heat gain is given by:

$$Q_{useful} = \dot{m} \, C_p (T_{out} - T_{in}) \tag{2}$$

where $\dot{m}$ represents the HTF mass flow rate, C_p is the specific heat of the HTF, and $(T_{out} - T_{in})$ is the temperature differences between the inlet and the outlet of the absorber tubes.

From the above two models, it is clear that to improve the thermal performance of the trapping mechanism many techniques could be followed. These techniques and how they are affecting the collector performance, their available modifications and design are discussed in the following sections.

Effect of the Collector Area

The incoming solar radiation is measured in W/m^2 and is absorbed by the absorber plate. Thus, the area of the absorber plate is directly proportional to the heat energy gain; i.e., larger plate area means overall higher energy absorbed. The material of the plate also plays an important role in energy gain because higher absorptance would give better yield. Normally, metals have higher thermal absorptance than non-metals; therefore, they are used for absorber plate or the tubes. Copper is one of the most popular metals used for absorber plate and tubes because of its thermal properties and ease of machining.

In designing of the collector receiver part, factors, like absorber tubes dimension, gap between tubes, flow direction and transient direction of the incoming solar radiation, play an important role in increasing the gain. If the design is based on relatively big diameter tubes, it would lead to increase the area but the traveled heat transfer depth will also increase, which eventually, will increase the thermal losses by convection. Thus, the overall heat transfer coefficient would be decreased, causing the flow rate to decrease too, which consequently leads to decrease in the kinetic energy.

On the other side, if the design is based on relatively small diameter tubes, the gap between tubes is decreased and this would start the interaction heat flow between the adjacent tubes resulting in an unsteady flow in the tubes. This means that the HTF would be absorbing heat and at the same time would be losing heat, thus making flow very complicated.

The parallel and series flow regimes inside the tubes are represented in Fig. 9. The gap between the tubes is a major factor of the collector interception area design. If the gap is increased, the spill out loss will increase too and so the collector efficiency will decrease. The main two flow regimes that are applied in the solar trapping collectors are the parallel flow, and in series flow, the second is used when high temperature at the outlet is required and the former regime is used when mass flow rate needs to be high.

Use of Finned Tubes

Design based on finned tubes is one of the methods used to increase the collector intercepting area and to improve the heat gain. This technique is widely applied in flat-plate collector applications; it helps in reducing the spill out losses for small tube designed collector approach. Increase in the overall absorption of the radiation would eventually increase the temperature of the tube and thus raise the water temperature.

Figure 10 shows the schematic of a solar collector with finned tubes. These fins intercept the radiation and simultaneously increase the heat energy absorbed by the absorber tubes, which is eventually transferred to the HTF, through conduction.

Effect of HTF Flow Rate

In the described mathematical model of Eq. (2), the mass flow rate, $\dot{m}$, is a control parameter. Once the collector parameters like the receiver area, absorber material properties, the incoming radiation, adjusted flow rate, and inlet temperature are deter-

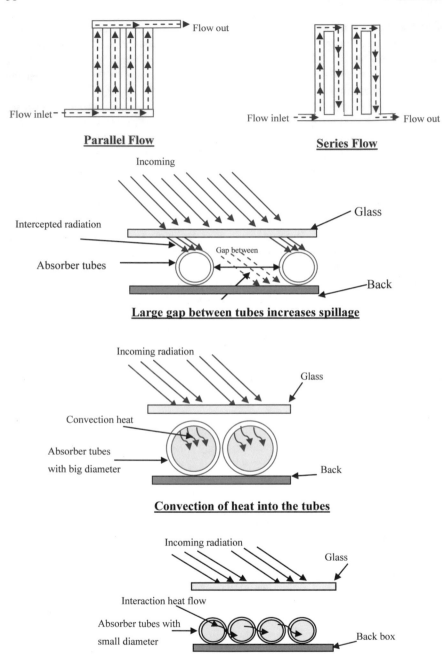

Fig. 9 Visual effects of various design parameters on the heat gain

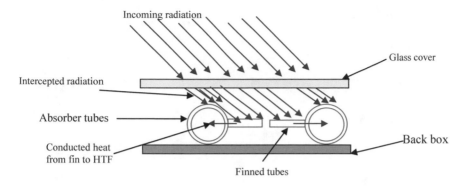

Fig. 10 Effect of finned tubes on heat gain of the collector

mined, the outlet temperate could be calculated. The relationship between mass flow rate and outlet temperature is a reverse relation; i.e., lower mass flow rate would increase the outlet temperature and vice versa.

Moreover, HTF mass flow rate and required outlet temperature play an important role in deciding the diameter of the tubes. Normally, the small tube diameters are more suitable for low mass flow rate and vice versa.

Effect of Number of Glazing Covers

Generally, a single glazing is used in flat-plate collectors, but the problem with such an arrangement is that it traps the heat in the trapping zone where the temperature of the air increases, and due to this high temperature at first trapping zone, the heat energy is lost to the environment at a larger rate. To avoid such energy losses from the glazing, a second or third glazing is used to avoid the thermal losses from the glazing cover. This technique is mainly used in low-temperature application collectors: increasing numbers of glazing covers boost the insulation of the absorber tubes and trapping mechanism as shown in Fig. 11.

Some of the heat is escaped out from the first glass cover; that is why, the second glass cover is used to reflect back this heat. The first trapping zone is hotter than the second one. This technique helps to decrease the temperature differences between the glass surface and ambient; so the second part of the first model is decreased which leads to increase the overall gain. However, the problem of this technique is that the incoming solar radiation is reduced by increasing the number of glass covers, thus reducing the input energy as well.

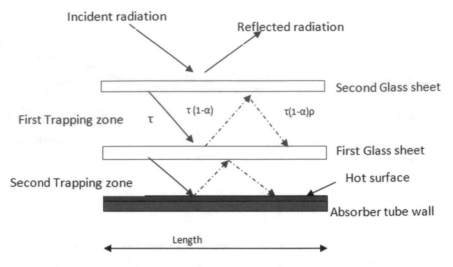

Fig. 11 Increase number of glazing covers

Effect of Material Properties

One of the most important factors that are used to improve the thermal performance of the trapping mechanism is the material properties. This includes the use of absorber with high absorbance property, i.e., high thermal energy absorption, glass covers with high transmittance with low conductivity and low emittance (normally low iron glass possess such properties), collector envelope box with good insulation to protect the back dissipation of heat.

Currently for tubing absorbers, other important factors to consider are the potential for corrosion, high-pressure durability, and mechanical properties. Copper alloy and steel alloy are the most applied material for making the tubes. Moreover, wide range of study is focused in improving the absorber conductivity by using special types of coating material called selective absorbers, with high absorption property, such as transparent conductive oxides (TCO) and multi-layer absorber $Al_2O_3/Mo/Al_2O_3$ [3].

For high-temperature application, different types of materials are used to improve the absorbance and receiver thermal ability under high-pressure environment. These materials include special metal–dielectric composite/cermets such as $W-Al_2O_3$ and $Mo-SiO_2$ cermets.

Use of Auxiliary Tracking Mechanism

Since sun position varies on the horizon during the daytime, the direction of the radiation falling on a horizontal surface would constantly be changing its incidence angle. The transient nature of the sun affects the amount of incoming radiation; that is why, instantaneous changes in the radiation level need to be measured. The most effective receiver aperture orientation is to directly face the sun, i.e., perpendicular to the radiation direction; otherwise, an optical loss known as cosine loss would reduce the impact of incoming radiation.

Auxiliary tracking mechanism is used in solar thermal applications to help in reducing cosine loss, especially for the collector, in high-temperature application category. Different types of tracking mechanisms are suggested and designed according to the required conditions. The most common tracking approaches are line tracking and point tracking models. The line tracking is applied in PTC and LFR, which is considered as an application required for relatively low temperature (>250 °C). When the point tracking is applied in PDS and SPT, it is considered as applications that are required for elevated temperature of up to 1000 °C.

Designing of tracking systems requires an accurate knowledge of the latitude, longitude, global coordinates of the collector location and local time zone. Mainly, hour angle, declination, azimuth, tilt, and collector orientation angles are the basic parameters that are involved in the design of tracking system.

Use of Concentrators

When the incoming amount of solar radiation is limited and insufficient to elevate the HTF temperature to a certain level, concentrators are used to increase the amount of solar radiation in a focus point, where the receiver is located. By this technique, the part of solar radiation in the above-described first model is increased, and thus, the total heat gain is increased and elevated temperature is achieved. A number of reflectors are used to focus the solar radiation on a point or line where the receiver is located, thus focusing, by many times, the light onto a point. This would increase the amount of radiation at any point and thus increase the incoming radiation to a level, which depends on the number of reflectors. This quantity is normally called the concentration ratio.

Many type of concentrators are designed and applied including mirrors, lenses, reflectors, or refractors, and many concentrator shapes are suggested including cylindrical, surfaces of revolution, continuous long unit and segmented unit. Concentrators are used in CSP system design. Detailed descriptions of different concentrator are previously shown in Figs. 4, 5, 6, and 7.

References

1. S.A. Kalogirou, *Solar Energy Engineering: Processes and Systems*, Academic Press (2013)
2. J.A. Duffie W.A. Beckman, Solar Engineering of Thermal Processes, vol. 3, (Wiley, New York, 2013)
3. M. Shimizu, M. Suzuki, F. Iguchi, H. Yugami, High-temperature solar selective absorbers using transparent conductive oxide coated metal. Energy Procedia. **57**, 418–426 (2014)

Air-to-Air Fixed Plate Energy Recovery Heat Exchangers for Building's HVAC Systems

Mohammad Shakir Nasif

The provision of fresh outdoor air to buildings has become a legislated standard practice in many countries around the world to improve indoor air quality. However, in HVAC systems the cost of conditioning the outdoor air is very high. For such systems, it is essential to utilize energy recovery devices to reduce this load. Therefore, air-to-air fixed plate heat exchanger is used in buildings where the room exhaust air is passed in one stream and ambient fresh air is passed in the other stream of the heat exchanger. This heat exchanger has two types which are: sensible heat exchanger which recovers only sensible heat and enthalpy heat exchanger which utilizes membrane and recovers both sensible and latent heat (dehumidify the supplied air). This heat exchanger will precondition the fresh air prior to supplying the air to an air conditioner resulting in substantial amount of energy saving.

Introduction

In the last few decades, building indoor air quality has become a great concern which is affected by volatile organic compounds, smoke, dust, and bacteria. As a result, many countries in the world have adopted new standards that specify higher indoor quality in buildings [1]. ASHRAE Standard 62 [2] defines acceptable indoor air quality as air in which there are no known contaminants at harmful concentrations and that a substantial majority of people exposed do not express dissatisfaction. Maintaining thermal comfort is not just desirable in assuring a productive work environment, but in many cases also has a direct effect on the health of the occupants.

Previous research showed that for an outdoor ventilation rate of 40%, the heating load would increase by around 30% as compared to no outdoor ventilation [3]. In

M. S. Nasif (✉)
Department of Mechanical Engineering, Universiti Teknologi PETRONAS, Perak, Malaysia
e-mail: mohammad.nasif@utp.edu.my

© The Author(s), under exclusive license to Springer Nature Singapore Pte Ltd. 2019   63
S. A. Sulaiman (ed.), *Sustainable Thermal Power Resources Through Future Engineering*, SpringerBriefs in Applied Sciences and Technology,
https://doi.org/10.1007/978-981-13-2968-5_5

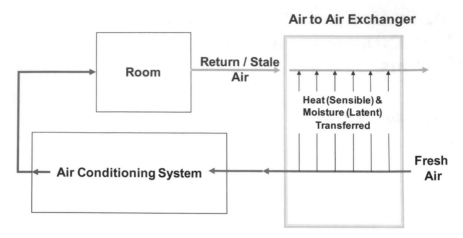

Fig. 1 Operating principle of air-to-air fixed plate exchanger

addition, the fresh air is a major source of moisture and will cause an increase in the latent load in the air-conditioning system. In fact, the latent load constitutes a large fraction of the total thermal load for an air-conditioned building. This latent load would increase when 100% fresh air is used in certain buildings, especially in a humid climate. Normally the water vapor in atmospheric air is small (some tens of grams per kilogram of fresh air), however, due to the high heat of vaporization, the latent load on an HVAC system constitutes a large fraction of the total thermal load. In addition, any excessive moisture content in a building must be dealt with by the HVAC system, as this moisture can lead to mold growth (particularly in humid climates) and indoor air quality problems [4].

There is a strong link between indoor air quality, temperature, humidity, and occupant productivity. However, meeting the minimum ventilation requirements may not be in the best interest of the building owner/operator, due to increased operating cost. Therefore, with energy costs rising, building owners and operators are looking for ways to conserve energy and lower utility bills using energy recovery systems.

Air-to-air fixed plate energy recovery system is one method proven to reduce energy consumption. This can be achieved by utilizing the room exhaust air to pre-cool or heat the fresh air before it enters the air-conditioning system as depicted in Fig. 1. Such systems reduce the HVAC operating cost.

Types of Fixed Plate Heat Exchangers

In general, there are two types of air-to-air energy recovery heat exchanger. The first type is the rotary wheels which require a motor to rotate the wheel. It does not provide 100% fresh air because of air carry over between the two streams. The second type

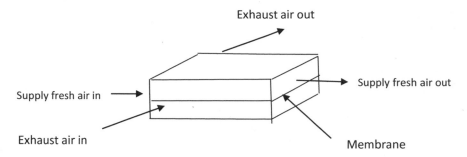

Fig. 2 Cross-flow air-to-air fixed plate exchanger

is fixed plates air-to-air heat exchanger. This exchanger is a static device that can be integrated into existing HVAC systems with minimum leakage between streams. It is less bulky than other energy recovery devices and has no complicated mechanisms and does not transfer dust and pollutants. Furthermore, it is cheap, does not require maintenance and external power to drive them, and it is simple to construct and safe to use in any ventilation system.

Energy recovery systems are classified into two categories, which are sensible and latent heat recovery. They are available as sensible heat exchangers and enthalpy heat exchangers, in which the designs are the same for both types of heat exchanger.

Sensible Fixed-Plate Heat Exchanger

The fixed-plate sensible heat exchanger consists of alternate layers of frames (which are known as flow paths). These layers form the supply and exhaust air stream passages, and these streams are separated by thin plates or plastic sheet.

The flow path allows the fresh and humid air to flow into the frames, whereas, the cold less humid air flows through the alternate frames which are laterally inverted with the fresh airframes. A thin plastic or metal sheet is used as a heat transfer surface between the two streams. Hence, only sensible heat will be transferred. The heat exchanger could be designed to provide cross-flow, or quasi-counter flow arrangement. Figure 2 shows cross-flow configuration of the heat exchanger.

Enthalpy Fixed Plate Heat Exchanger

This heat exchanger was developed with the first prototype tested unit in the year 1970 [5] and was developed due to the increase in demand to recover latent heat and to overcome the disadvantages and limitations of other energy recovery devices. In fact, it is an enhanced version of the sensible fixed-plate heat exchanger where

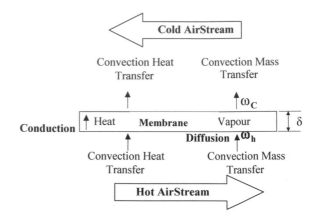

Fig. 3 Heat and moisture transfer across the membrane

conventional plates are replaced by a permeable material (such as membrane) that is able to transfer both heat and moisture.

The operation of membrane-based enthalpy heat exchangers is based on the second law of thermodynamics which states that energy always transfers from a region of high temperature to one of a low temperature. This law can be extended to say that mass transfer always occurs from region of high vapor pressure to one of low vapor pressure through a permeable material.

In fact, both heat and mass transfer phenomena influence membrane-based enthalpy energy recovery systems, where the ambient hot and humid supply air is passed over one side of a membrane heat exchanger and on the other stream the cold and less humid air is passed. Due to the gradient in the temperature and vapor pressure concentration, heat and moisture are transferred across the membrane surface.

The heat and moisture are transported from the hot and humid stream to the membrane surface by convection, followed by conduction of heat and diffusion of moisture through the porous membrane surface and by convection from the membrane surface to the cold and less humid stream, causing a decrease in temperature and humidity of the supply air stream before it enters the evaporator unit, hence both sensible and latent energy are recovered as shown in Fig. 3.

Heat Exchanger Performance

The performance of the heat exchanger is usually evaluated by calculating the heat exchanger sensible, latent and total effectiveness, where, the effectiveness represents the ratio between the actual heat transfer rate and the maximum heat transfer rate. The sensible heat transfer represents the amount of heat transferred due to temperature difference; latent heat transfer represents the amount of latent heat transferred due to moisture content difference (vapor pressure difference); and total heat transfer represents the amount of total heat transferred due to the enthalpy difference between the heat exchanger streams. The heat exchanger effectiveness can be determined through experimental measurements and mathematically by using Effectiveness-NTU method.

Effectiveness from Experimental Measurements

Through measuring the air inlet and outlet conditions (temperature and humidity), the general form of the sensible effectiveness is:

$$\varepsilon_S = \frac{\dot{m}_s C_p (T_{hi} - T_{ho})}{\dot{m}_{min} C_p (T_{hi} - T_{ci})} = \frac{\dot{m}_e C_p (T_{co} - T_{ci})}{\dot{m}_{min} C_p (T_{hi} - T_{ci})} \tag{1}$$

Analogous to the sensible effectiveness, the latent effectiveness is calculated based on the amount of moisture transferred and is represented by:

$$\varepsilon_L = \frac{\dot{m}_s h_{fg} (\omega_{hi} - \omega_{ho})}{\dot{m}_{min} h_{fg} (\omega_{hi} - \omega_{ci})} = \frac{\dot{m}_e h_{fg} (\omega_{co} - \omega_{ci})}{\dot{m}_{min} h_{fg} (\omega_{hi} - \omega_{ci})} \tag{2}$$

and the total effectiveness is expressed as:

$$\varepsilon_{tot} = \frac{\dot{m}_s (H_{hi} - H_{ho})}{\dot{m}_{min} (H_{hi} - H_{ci})} = \frac{\dot{m}_e (H_{co} - H_{ci})}{\dot{m}_{min} (H_{hi} - H_{ci})} \tag{3}$$

where $\dot{m}$ is air mass flow rate, T_{hi}, ω_{hi}, and H_{hi} are ambient air inlet temperature, humidity ratio, and enthalpy, T_{ho}, ω_{ho}, and H_{ho} are air outlet temperature, humidity ratio, and enthalpy which enters the cooling coil, T_{ci}, ω_{ci}, and H_{ci} are room exhaust air temperature, humidity ratio, and enthalpy, and h_{fg} represents air enthalpy of evaporation.

From the above equations, it is clear that the ideal energy transfer would move 100% of the energy difference between the stream with the higher temperature and humidity to the stream with lower temperature and humidity. From the above equations too, the sensible and enthalpy heat exchanger performances are determined.

Effectiveness-NTU Method

To determine the sensible and latent effectiveness, two mathematical models must be used. The sensible model is detailed as follows, in which the sensible heat transfer model is based on Nusselt number and Reynolds number correlations for each air channel.

When the Reynolds number is less than 2000, Hausen's correlation is recommended for laminar flow in ducts under uniform heat flux [6]. The correlation includes flow developing effect in heat exchangers:

$$\mathrm{Nu} = 8.235 + \frac{0.0068 \left(d_{\mathrm{hy}}/L \right) \mathrm{Re} \, \mathrm{Pr}}{1 \, + \, 0.04 \big[\left(d_{\mathrm{hy}}/L \right) \mathrm{Re} \, \mathrm{Pr} \big]^{2/3}} \tag{4}$$

where d_{hy} is the hydraulic diameter, L is the developed length. When the flow is turbulent the following correlation is used [6]:

$$\mathrm{Nu} \, = \, 0.036 \, \mathrm{Re}^{0.8} \, \mathrm{Pr}^{\frac{1}{3}} \left(\frac{d_{\mathrm{hy}}}{L} \right)^{0.055} \tag{5}$$

The Reynolds number of a flow is calculated as:

$$\mathrm{Re} \, = \, \frac{V \, d_{\mathrm{hy}}}{\nu} \tag{6}$$

where V is the air velocity and ν is the kinematic viscosity. The Prandtl number, Pr, in Eq. (5) represents the ratio of fluid molecular diffusivity and molecular diffusivity of heat. By determining the Nusselt number from correlations above, the convective heat transfer coefficient (h_{heat}) is calculated:

$$h_{\mathrm{heat}} \, = \, \frac{\mathrm{Nu}\lambda}{d_{\mathrm{hy}}} \tag{7}$$

where the Nusselt number, Nu, represents the convective heat transfer through a fluid layer relative to the conduction across the same layer.

The total number of transfer units is:

$$\mathrm{NTU}_s \, = \, \frac{A_{\mathrm{ht}} U_s}{C_{\mathrm{min}}} \tag{8}$$

where A_{ht} is the area of heat transfer, and the heat capacity of the stream is:

$$C \, = \, \dot{m} C_p \tag{9}$$

where $\dot{m}$ is air mass flow rate, and C_p is air specific heat. The general form of overall sensible heat transfer coefficient, U_s, is:

$$U_s \, = \, \left[\frac{1}{h_{h,\,\mathrm{heat}}} + \frac{\delta}{k_{\mathrm{ther}}} + \frac{1}{h_{c,\,\mathrm{heat}}} \right]^{-1} \tag{10}$$

The thermal resistance (δ/k_{ther}) represents the conduction resistance of this heat transfer surface sheet and is a function of the material type and thickness. However, this term is very small due to the small thickness of the heat transfer film. The other two terms are the convective heat transfer resistance in the two streams, and these terms contribute the largest portion of the sensible heat transfer resistance (U_s).

The sensible effectiveness for cross-flow is:

$$\varepsilon_{s,\,\text{cross}} = 1 - \exp\left\{\frac{\text{NTU}_s^{0.22}}{\frac{C_{\min}}{C_{\max}}}\left[\exp\left(-\frac{C_{\min}}{C_{\max}}\text{NTU}_s^{0.78}\right) - 1\right]\right\} \quad (11)$$

Latent heat transfer model that simulates moisture transfer can be developed using the convective mass transfer Sherwood correlation. The convective mass transfer coefficient can be obtained using the Chilton-Colburn analogy [7]:

$$\text{Sh} = \text{Nu.Le}^{-\frac{1}{3}} \quad (12)$$

where Sherwood number (Sh) represents the convective mass transfer through a fluid layer relative to the mass diffusion across the same layer. Sherwood number is represented by

$$\text{Sh} = \frac{h_{\text{mass}}d_{\text{hy}}}{D_{\text{va}}} \quad (13)$$

The Nusselt and Sherwood numbers represent the effectiveness of convective heat and mass convection at the surface, respectively.

The Lewis number represents the relative magnitudes of heat and mass diffusion in the thermal and concentration boundary layers and is defined as:

$$\text{Le} = \frac{K_{\text{air}}}{C_{p\,\text{moist air}}\,D_{\text{water}-\text{air}}\,\rho_{\text{moist air}}} \quad (14)$$

By substituting Eq. (12) for Nusselt number and Eq. (14) into Eq. (13), the convective mass transfer coefficient (h_{mass}) can be represented by:

$$h_{\text{mass}} = \frac{h_{\text{heat}}}{C_{\text{pa}}}\text{Le}^{-\frac{1}{3}} \quad (15)$$

During the moisture transfer measurement in this study steam was injected into the hot stream in order to increase the vapor pressure, and thus, the Lewis number is typically 0.81 [8]. The number of latent energy transfer units (NTU$_L$) is:

$$\text{NTU}_L = \frac{A_{\text{ht}}U_L}{\dot{m}_{\min}} \quad (16)$$

The overall mass transfer coefficient (U_L) is given by:

$$U_L = \left[\frac{1}{h_{h,\text{mass}}} + R_{\text{paper}} + \frac{1}{h_{c,\text{mass}}}\right]^{-1} \quad (17)$$

The moisture resistance of the porous material is R_{paper}, and the other two terms are the convective mass transfer resistance of each flow stream. Researchers found that contrast to the membrane conduction thermal resistance which is almost constant, the membrane surface moisture transfer resistance varies with the humidity ratio difference across the paper surface [8].

In general, membranes have been classified into three categories according to the membrane moisture resistance variation under different humidity conditions. The first type is a membrane with constant moisture transfer resistance, where researchers found that the moisture resistance is independent of the air humidity. The second is a membrane with moisture resistance increasing in line with increasing the humidity. For this type of membrane, researchers found that the membrane moisture uptake decreases when the humidity increases. The third type is where the membrane mois-ture uptake increases when the humidity increases and subsequently the moisture resistance decreases with increasing the humidity [9]. To calculate the cross-flow latent performance, the latent effectiveness is represented as:

$$\varepsilon_{L,\,cross} = 1 - \exp\left\{\frac{NTU_L^{0.22}}{\frac{\dot{m}_{min}}{\dot{m}_{max}}}\left[\exp\left(-\frac{\dot{m}_{min}}{\dot{m}_{max}}NTU_L^{0.78}\right) - 1\right]\right\} \qquad (18)$$

Energy Saving

Several studies were performed to investigate the energy saving fixed plate heat exchanger can provide. Nasif et al. [10] studied the annual energy consumption of an air conditioner coupled with an enthalpy/membrane heat exchanger and compared with a conventional air-conditioning cycle using in-house modified HPRate software. The heat exchanger effectiveness is used as thermal performance indicator and incor-porated in the modified software. Energy analysis showed that an air-conditioning system coupled with a membrane heat exchanger consumes less energy than a con-ventional air-conditioning system in hot and humid climates where the latent load is high. It has been shown that in humid climate a saving of up to 8% in annual energy consumption can be achieved when membrane heat exchanger is used instead of a conventional HVAC system.

Nasif [11] investigated the performance of air-to-air fixed plate energy recovery heat exchanger utilizing porous paper and Mylar film as the heat and moisture transfer media used in ventilation energy recovery systems. This performance is represented by the heat exchanger sensible and latent effectiveness. A simplified air-conditioning system which is represented by cooling coil that incorporates air-to-air fixed plate heat exchanger to cool office space is developed under humid inlet air conditions. Energy analysis for tropical climate shows that utilizing paper surface heat exchanger in a standard air-conditioning system will lead to 78% energy saving as compared with utilizing Mylar plastic film which recovers only sensible heat.

These results show the significant contribution of the enthalpy heat exchanger in reducing the latent load in hot and humid climates, and the substantial energy savings achieved in comparison with conventional air-conditioning systems. In addition, enthalpy heat exchanger has the advantage of providing 100% fresh air.

Summary

Air-to-air fixed plate energy recovery exchanger is a static device which recovers substantial energy whilst providing 100% fresh air supply through HVAC systems. The performance of the heat exchanger can be calculated through determining the effectiveness. The effectiveness also can be predicted by using effectiveness-NTU method; however, the latent effectiveness requires determining the membrane moisture transfer resistance which its value changes according to the inlet air humidity.

Air-to-air fixed plate exchanger is a simple static device which requires less maintenance and can be easily integrated into HVAC system as compared to other energy recovery systems. Past research showed that air-to-air fixed plate membrane energy recovery performed very well and recovered high energy in humid climate as compared to moderate and dry climate conditions.

References

1. ASHRAE, ASHRAE Handbook—fundamentals, American society of heating, (Refrigerating and Air-Conditioning Engineers Inc., Atlanta, 1999)
2. ASHRAE Standard 62, Ventilation for acceptable indoor air quality, American society of heating, (Refrigeration and Air Conditioning Engineers, Inc., Atlanta, 2001)
3. A. Nguyen, Y. Kim, Y. Shin, Experimental study of sensible heat recovery of heat pump during heating and ventilation. Int. J. Refrig **28**, 242–252 (2005)
4. H. Crowther, Enthalpy Wheels for Energy Recovery. J. Heating Piping Air Conditioning Eng. **73**, 39–45 (2001)
5. A.A. Field, Heat recovery from air systems. J. Heating Piping Air Conditioning **72**, 73–78 (1975)
6. J.P. Holman, *Heat Transfer* (McGraw-Hill Inc, USA, 1997)
7. W.M. Kays, M.E. Crawford, *Convective heat and mass transfer*, 4th edn. (McGraw-Hill International Edition, Singapore, 2005)
8. M.S. Nasif, R. Al-Waked, M. Behnia, G. Morrison, Air to air fixed plate enthalpy heat exchanger, performance variation and energy analysis. J. Mech. Sci. Technol. **27**(11), 3541–3551 (2013)
9. M.S. Nasif, R.F. Al-Waked, M. Behnia, G. Morrison, Modeling of air to air enthalpy heat exchanger. Heat Trans. Eng. **33**(12), 1010–1023 (2012)
10. M. Nasif, R. Al-Waked, G. Morrison, M. Behnia, Membrane heat exchanger in HVAC energy recovery systems, systems energy analysis. Energy Build. **42**(10), 1833–1840 (2010)
11. M.S. Nasif, Effect of Utilizing Different Permeable Material in Air-To-Air Fixed Plate Energy Recovery Heat Exchanger on Energy Saving. ARPN J. Eng. Appl. Sci. **10**, 10153–10158 (2015)

Strategy for Cooling of Computing Data Center Facilities

Shaharin Anwar Sulaiman

This chapter reports a study on the optimum cooling energy required in operating a campus grid outside normal working hours. The main issue involved in this study includes assurance of acceptable operating air temperature for the computers in the data center facility (laboratories) when the air-conditioning system is turned off. At the same time, the study looks into minimizing the cooling energy required to overcome the heat dissipated from the CPUs, and thus ensuring conservation of energy.

Introduction

Energy consumption in Malaysia has increased quite steadily in the last decade due to its high economic growth and increase in the standard living of households. However, energy is becoming costlier due to the increase in global demand and drop in production of fossil-based fuel mainly from the oil and gas sector. On the environmental aspect, global warming resulted from greenhouse gas emission through combustion of fuels has received significant attention from the world. Thus, efficient energy usage and significant reduction in the released emission are therefore required. Cooling energy in Malaysia accounts for 42% of total electricity energy consumption for commercial buildings and 30% of residential buildings [1]. Numerous efforts on optimization of cooling energy in buildings' air-conditioning systems were reported. Among others, Sulaiman and Hassan [2] demonstrated that cooling energy can be optimized by cutting off the supply of cool air to unoccupied rooms by proper scheduling of the usage of rooms. In another work, it was shown that significant reduction in cooling energy can be achieved by delaying the start-up of AHU or

S. A. Sulaiman (✉)
Universiti Teknologi PETRONAS, Perak, Malaysia
e-mail: shaharin@utp.edu.my

© The Author(s), under exclusive license to Springer Nature Singapore Pte Ltd. 2019   73
S. A. Sulaiman (ed.), *Sustainable Thermal Power Resources Through Future Engineering*, SpringerBriefs in Applied Sciences and Technology,
https://doi.org/10.1007/978-981-13-2968-5_6

chilled water supply for building areas which are not facing the sunrise and also by shutting down the system 30–60 min earlier than usual [3]. The role of set-point temperature of the air is also important as the total building electricity can be reduced by 3% for every °C rise in the indoor air temperature [4]. Intelligent computer programs were also proven in achieving significant savings in cooling energy, for example, in the work by Lee and Braun [5] who demonstrated 30% reductions in peak cooling loads through a demand-limiting strategy that employed an inverse building model trained with field measurements.

In the case of computer data center facilities, reduction in cooling energy can help playing roles in overcoming these problems since the energy involved is not negligible due to increase in interest and usage of computing systems in the society. During the day, air conditioners are mainly intended to serve human or better referred to as occupants. Here, the purpose of cooling is mainly to achieve satisfactory thermal comfort level, which mainly covers air temperature, humidity, air motion, and radiation. Furthermore, during the day, the source for heat gain is mainly from the sun via conduction through walls and radiation through glazing. In this case, the effect of heat dissipation from computers would be relatively negligible. The situation changes after sunset when the main source of heat gain is absent. Furthermore, offices and laboratories would normally be unoccupied after the end of working hours. With no occupants to serve, the room air temperature can be increased to a higher value is tolerable to computers, and thus this creates an opportunity for reduction in cooling energy. However, it is uncertain of whether air conditioners would still be required at night or otherwise. There could also be solutions in between the two options. Hence, a study is required to understand better the nature of heat transfer inside the computer data center facility at outside working hours prior to deciding the suitable cooling mechanism.

The objective of this work is to assess the optimum cooling energy required in computer laboratories when they are functioning as part of a campus grid network for computing. The study is conducted by assessing the room air temperature entering the inlet air intake of computers in the room at different time and conditions as well as different cooling configurations. The power consumed by the computing equipment and also the total power consumed by the computing facility are also measured in order to assess the facility's power usage effectiveness (PUE). The definition for PUE can be found in various reports, e.g., by Fontecchio [6]. In addition, the green power usage effectiveness (GPUE), as proposed by Hrafnsson [7], can also be determined to measure the facility's carbon footprint per usable energy (kWh). The outcome of the study can be used to properly judge on the most suitable method or system for optimum cooling of the computer data center facility throughout a day in tropical countries like Malaysia.

Method of Study

The study was performed in a few stages onto selected computer laboratories in Universiti Teknologi Petronas (UTP). The first stage involved measurements of air temperature and humidity at the inlet of computers in the data center facilities. The purpose of the measurements was to gauge the limit of conditions at which air flowing into the computer would be acceptable when the computers served as part of the campus computing network. In the second stage, the PUE was determined by continuous measurements of thermal and electrical power in a selected computer laboratory, under various conditions. This would enable comparison of the effectiveness of the present cooling system with benchmark values. Lastly, analyses of results from the first two stages were used to configure, at preliminary level, suitable cooling mechanisms for the campus computing network.

Descriptions of the Buildings

The study was conducted in selected computer clusters located within the academic complex of a university campus in Malaysia, located in Tronoh, 200 km to the north of Kuala Lumpur. The average ground elevation of the place is 62 m above the sea level. The average outdoor dry bulb temperature and wet bulb temperature were 32.5 and 26.9 °C, respectively. The campus academic complex comprises 16 buildings (blocks) that were constructed next to each other, as shown in Fig. 1, and they consist of classrooms, offices, laboratories, and computer rooms. Each of the buildings has four floors except for Blocks 3 and 15 which have three floors. The aspect ratio of the typical building is 3.1, and the total air-conditioned floor area is approximately 4833 m^2 for each building. In general, air-conditioning is supplied for 12 h per day, typically between 7.30 am and 7.00 pm from Mondays to Fridays. During the weekends, air-conditioning system in the buildings is supplied by demand, usually for selected laboratories and lecturers' offices. Each floor has two air handling units (AHUs). The main controls of the centralized air-conditioning systems, including the start-up and shutdown time of the AHU's, are remotely operated from the control room located in the administration building within to the academic complex. Each of the air ducts in the buildings is equipped with a variable air volume (VAV) system for efficient control of energy by varying the supply rate of cool air based on the response by temperature sensors. Descriptions of the VAV system can be found in various references, e.g., in Pita [8]. The buildings also have other energy saving mechanisms such as the variable speed drive for the blower's motor of the AHU, heat recovery wheels, and motorized valve for modulation of chilled water supply.

At times, the computers in selected computer clusters were loaded with a finite-difference time-domain (FDTD) simulation of electromagnetic wave propagation using a simulation grid, which was large enough to cause disk thrashing. The purpose of such loading was to mimic real grid networking conditions, which were expected

Fig. 1 Aerial view of the academic complex

to generate high amount of heats and consequently causing the room air temperature to increase. For the computers in the laboratories, a simulation grid of 6000 by 6000 would be enough to cause disk thrashing.

Measurements of Air Temperatures and Humidity

The computer cluster in Room 02-01-02 in Block 02 was considered in the study because it was not utilized by students or lecturers during the period of study. The height of the room was 3.1 m and the floor was of 11.3 m by 20.3 m. The room was equipped with 40 desktop computers.

Air temperatures were recorded at various points in the room, mainly at the air inlets and air outlets of selected computers. Type K thermocouples were used for the temperature measurements. The thermocouples were connected to data loggers and computer to enable continuous recording at an interval of 10 min. A portable CO_2/temperature monitor (Telaire 7001) was used to monitor the relative humidity. The data was stored and analyzed using HOBOware Lite software. The outdoor conditions of air were also recorded using a commercial weather station device. All measurements were automated, and therefore human intervention was minimized. Shown in Fig. 2 is the general layout of the room, computers, humidity sensor, and thermocouples. Six pairs of thermocouples were used to measure the inlet and outlet air temperature entering and leaving each CPU. In this case, each thermocouple was positioned to hang 60 mm away from the inlet or outlet face and hanging 30 mm from the top edge of the CPU. The CPUs were mounted flat (not standing). Two other thermocouples were used to measure the room air temperature, and they were mounted hanging 710 mm from the floor.

The measurements were conducted during the weekend starting from Friday at 5.00 pm and were terminated at 8.00 am on the following Monday. The period (weekend) was chosen as it could cover both night time and weekend, i.e., outside working hours, during which the campus computing network was expected to be highly active. The conditions of test were varied, among others without air conditioners, with air supplier but without chilled water supply to the air handling units and with cool air supply at night without supply of chilled water to the air handling unit (AHU). In addition to the computer cluster, the AHU also served other neighboring rooms. The chilled water supply for the AHU was delivered from district cooling plant in the campus. In the study, it was noted that the computers' operating limit would be:

- Maximum intake air temperature of 28 °C,
- Maximum relative humidity of 80% [9].

Measurements of PUE and GPUE

To measure the power usage effectiveness (PUE) the following measurements were made:

1. Thermal energy consumption for cooling of the computer room,
2. Electrical power consumption for cooling of the computer room,
3. Electrical power consumption for operation of CPUs that served the grid computer network.

However, at the time of the study, the only place in UTP where thermal energy consumption of a room could be measured was Level 1 of Block 17, i.e., female-toilet wing. One available computer cluster within this air-conditioning zone was Room 17-01-15. It had a floor area of 11.3–16 m with a ceiling height of 3.1 m. Within the room was a technician's room located at a corner, which took a floor space of 9 m^2. There were 27 PCs in the room, and they were distributed on five round tables.

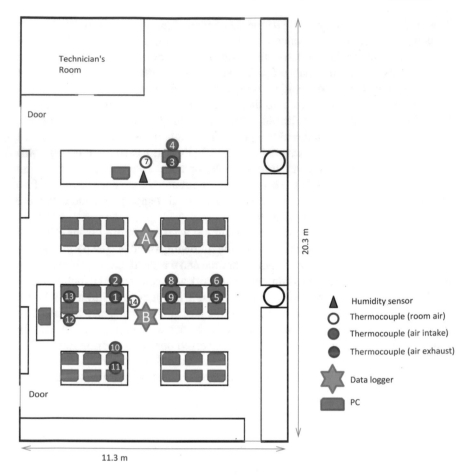

Fig. 2 General layout of the measurement systems in computer cluster Room 02-01-02 in Block 02

In the study, the supply of cool air was configured so that only the computer cluster was cooled while other rooms were not air-conditioned. The configuration was made with the assistance of the maintenance personnel by shutting the motorized dampers that served the other rooms. The cooling energy for the computer cluster was determined by manipulating the chilled water return and supply temperature and flow rate by:

$$\text{CE} = \rho \times \text{SG} \times C_p \times (T_{\text{CHWR}} - T_{\text{CHWS}}) \times \dot{V} \times \frac{1\ \text{hr}}{3600\ \text{s}} \quad (1)$$

where CE is the cooling load (kW), ρ is the density of water (1000 kg/m^3), SG is the specific gravity of water (SG = 1), Cp is the specific heat of water (4.23 kJ/kg. °C),

T_{CHWR} is the chilled water return temperature (°C), T_{CHWS} is the chilled water supply temperature (°C), and is the chilled water return volume flow rate (m³/hr).

Temperatures of supply and return chilled water were measured by temperature sensors, which were mounted in the pipelines within the plant room. The temperature sensor, R.S. Components LM35 CZ (317-960), had an operating temperature range of −40 °C to +110 °C with a linear output of 10 mV/°C. The accuracy of this instrument was ±1%. The chilled water flow rate was measured by using Maghant electromagnetic flow meter, which was manufactured by Endless+Hauser. The instrument was installed in the chilled water return pipe (beside the balancing valve) in the plant room. The flow meter could be operated under process temperature of between −20 and 120 °C and pressure of up 16 bar. The accuracy of this flow meter is ±2% at a point close to the measuring electrode.

Results and Discussions

The results are presented based on observations from the measurements conducted in the two computer clusters, each in Blocks 02 and 17. The involvement of the latter was mainly due to the inability to measure thermal energy in other rooms, which was discovered at a later stage of the project. In Block 02, the temperature and humidity behavior was the primary interest, whereas in Block 17 the study was intended to determine the PUE.

Air Temperatures and Humidity

Shown in Fig. 3 is the variation of air temperatures at different points in the computer cluster (Room 02-01-02) in Block 02. The measurements were conducted on June 9 and 10, 2012, (weekend) when it was not occupied and the air-conditioning system was turned off. The measurement started at midnight and was continuous for the next 48 h. In general, it is shown in Fig. 3 that the temperatures were at between 21 and 25 °C, which were below the set limit (28 °C). The trends of temperatures on both days were nearly consistent with very little temporal variations.

The fluctuations displayed in the graph in Fig. 3 are most probably the effect of heating due to the presence of sun which caused radiation heat transfer and also conduction temperature due to difference in between outdoor and indoor temperatures. The temperatures were different to each other, and this was most probably because different locations would expect to receive different heat gain from example solar radiation. The sudden drop in temperatures at 12 noon on June 10 was most probably due to accidental start of the AHU which brought down the room's air temperature.

In Fig. 4, similar measurements to that in Fig. 3 were conducted in the same room but with the computers turned on. In addition, the computers were running a program to mimic grid computing, and therefore it would generate more heat for the

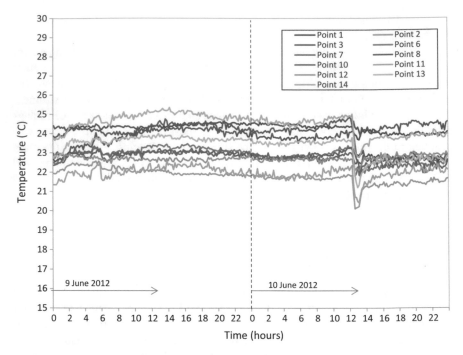

Fig. 3 Variation of air temperatures at different points in the computer cluster of Block 02 as measured on June 9 and 10, 2012, with the entire PC turned off

study. The measurements were conducted on April 7 and 8, 2012 (weekend), which started at midnight. With the computer turned on and busy running program, the heat generated can be harmful to the computers. The air temperatures are shown to be between 20 and 28 °C, which were still within the acceptable range. Furthermore, it is shown in Fig. 4 that the temperature of the air at most of the CPU inlets was lower than 23.5 °C. Although high temperatures were observed for the air leaving the CPUs, they were still within acceptable values. Hence, it can be suggested that when operating during the weekend, supply of cool air by the air conditioners may not be required in ensuring healthy operating condition of the computers.

Power Usage Effectiveness (PUE)

Shown in Fig. 5 is the variation of thermal power consumption or rate of cooling energy (CE) in the computer cluster of Block 17 with time for 24 h. The test was conducted during a weekend. The thermal power was determined using Eq. (1) with the chilled water temperatures and water flow rates obtained from the control room.

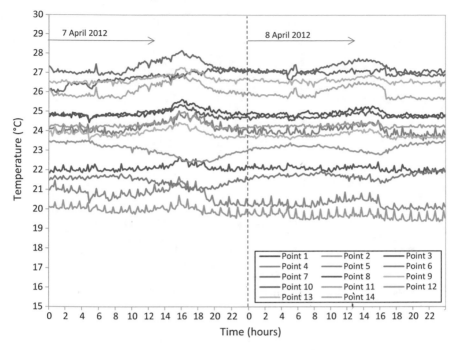

Fig. 4 Variation of air temperatures at different points in the computer cluster of Block 02 as measured on April 7–8, 2012, with the PC turned on and running some programs

At midnight, it is shown in Fig. 5 that the thermal power consumption was decreasing; this was because the chillers in the district cooling system plant were just turned off.

At about 6.30 am, the chillers in the district cooling plant and AHUs were turned on. In the case of Room 17- 01-15, although the AHU was intentionally turned on all the time, it was only at this time that the AHU's motorized valve was turned on, and thus allowing cooling by chilled water to take place. As a result, there was a sharp increase in the thermal power consumption since the control system responded to the room temperatures (see Fig. 3) which were relatively higher than the set-point temperature (°C). The sudden increase in the thermal power consumption at this time was due to the relatively large temperature difference as if there was a large deficit in cooling at that time. Once the set-point temperature was achieved, the cooling energy required to maintain the temperature would be less as shown in Fig. 5. It is unclear why there was a spike in the graph at about 10.30 pm. A possible explanation could be that a staff entered the room to check and could have probably left the door open for quite some time.

Calculation of the power usage effectiveness (PUE) showed that the trend was similar to the graph in Fig. 5. Shown in Fig. 6 is the variation of PUE at night time approximately the period when the chillers in the district cooling plant were turned off. This would be the period when grid network could be exposed to high-

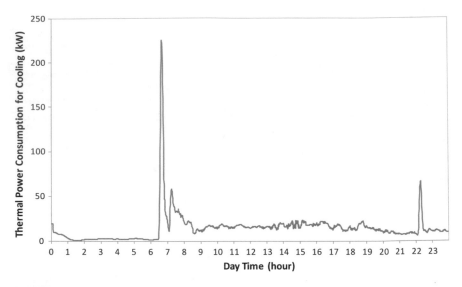

Fig. 5 Variation of thermal power consumption in the computer cluster of Block 17

temperature risk. Obviously, Fig. 6 shows that the PUE was at an average value of about 1.6, which could be considered as acceptable for a data center; an ideal value would be 1.0. It must be noted that the air-conditioning system was originally designed to cater for human occupation. If the room is dedicated for placement of computers only, the space would be made smaller, and hence the cooling energy would be very much smaller, and the electrical energy for the computers would be a few times higher, and thus the PUE could be further reduced.

Figure 7 shows the PUE variation during regular office hours (8.00 am to 6.00 pm). As anticipated, it is shown in the PUE was high at an average of about 3.5, which was more than two times larger than at night when the chillers were turned off. As mentioned earlier, this was due to the system design which was catered for human occupation, and therefore the set-point temperature was fixed at 24 °C. Furthermore, during the day, the presence of solar radiation would be significant, especially considering that the walls were made of glass that allowed sunlight to penetrate into the room. The results in Figs. 6 and 7 clearly show how significant saving can be achieved by allowing parallel grid computing at night. The potential to improve further is high.

Summary

In this study, it was indicative that operating the grid computing during the weekends and nights when the air conditioners were turned on would be harmless for the computers. The energy saving by not turning on the air conditioners during this period would be significant. It must, however, be noted that the study was still pre-

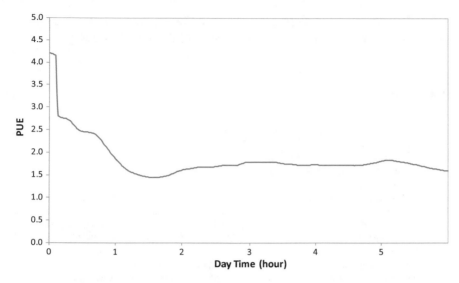

Fig. 6 Variation of PUE in the computer cluster of Block 17 during night time after chillers turned off

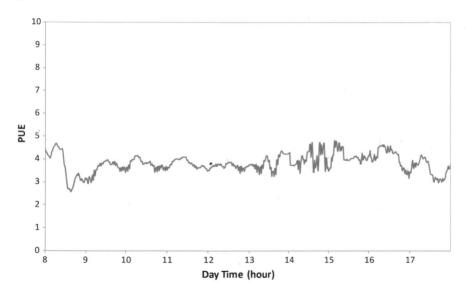

Fig. 7 Variation of PUE in computer cluster of Block 17 during daytime

liminary, and therefore the results were just early indicators of the expected trends. For example, the rooms of study may vary in size, and their orientations with respect to the sun position varied significantly due to the architectural design concept; see Fig. 1. Furthermore, provisions of computers per room and also their specifications varied. Therefore, a more thorough study would be required in order to acquire a more comprehensive and accurate understanding of the overall situation.

References

1. R. Saidur, M. Hasanuzzaman, M.M. Hasan, H.H. Masjuki, Overall thermal transfer value of residential buildings in Malaysia. J. Appl. Sci. **9**, 2130–2136 (2009)
2. S.A. Sulaiman, A.H. Hassan, Analysis of annual cooling energy requirements for glazed academic buildings, in *2nd International Conference on Plant and Equipment Reliability Kuala Lumpur*, (Malaysia, 2010)
3. S.A. Sulaiman, F.H. Tamby, M.F. Khamidi, *Reduction of cooling energy for highly glazed buildings in malaysia through scheduled start-up and shutdown of air-conditioners, presented at the 5th international ege energy symposium and exhibition (IEESE-5)* (Denizli, Turkey, 2010)
4. J.C. Lam, Energy analysis of commercial buildings in subtropical climates. Build. Environ. **35**, 19–26 (2000)
5. K.-H. Lee, J.E. Braun, *Reducing peak cooling loads through model-based control of zone temperature setpoints presented at the american control conference* (IEEE, New York, USA, 2007)
6. M. Fontecchio, Power usage effectiveness (PUE), search data center, TechTarget, April 2009, http://searchdatacenter.techtarget.com/definition/power-usage-effectiveness-PUE, Accessed on 4 October 2012
7. E. Hrafnsson, GPUE—Green power usage effectiveness, Greenqloud: News and Blog, 7 March 2012, http://blog.greenqloud.com/greenpowerusageeffectiveness-gpue/ Accessed on 4 October 2012
8. E.G. Pita, *Air Conditioning Principles and Systems: An Energy Approach*, 4th ed. (Prentice Hall, 2001)
9. C. Borysowich, Humidity and Computers, Blog: Toolbox for IT, 15 February 2005, http://it.toolbox.com/blogs/enterprise-solutions/humidity-and-computers-3158, Accessed on 4 October 2012

Printed in the United States
By Bookmasters